EcoProduction

Environmental Issues in Logistics and Manufacturing

Series editor

Paulina Golinska, Poznan, Poland

About the Series

The EcoProduction Series is a forum for presenting emerging environmental issues in Logistics and Manufacturing. Its main objective is a multidisciplinary approach to link the scientific activities in various manufacturing and logistics fields with the sustainability research. It encompasses topical monographs and selected conference proceedings, authored or edited by leading experts as well as by promising young scientists. The Series aims to provide the impulse for new ideas by reporting on the state-of-the-art and motivating for the future development of sustainable manufacturing systems, environmentally conscious operations management and reverse or closed loop logistics.

It aims to bring together academic, industry and government personnel from various countries to present and discuss the challenges for implementation of sustainable policy in the field of production and logistics.

More information about this series at http://www.springer.com/series/10152

Pawel Zajac

Evaluation Method of Energy Consumption in Logistic Warehouse Systems

Springer

Pawel Zajac
Faculty of Mechanical Engineering
Wrocław University of Technology
Wrocław
Poland

ISSN 2193-4614 ISSN 2193-4622 (electronic)
EcoProduction
ISBN 978-3-319-22043-7 ISBN 978-3-319-22044-4 (eBook)
DOI 10.1007/978-3-319-22044-4

Library of Congress Control Number: 2015948152

Springer Cham Heidelberg New York Dordrecht London

Printed on acid-free paper

Springer International Publishing AG Switzerland is part of Springer Science+Business Media
(www.springer.com)

This book is dedicated to my beloved wife,
Agnes, children Elizabeth and Charles,
who have always supported me
and were forgiving

Acknowledgments

The book "Evaluation Method of Energy Consumption in Logistics Warehouse Systems" was reviewed independently by two investigators Professors: Professor Kazimierz Wojs Phd., D.Sc., Wroclaw University of Technology, Wroclaw, Poland and Professor Jerzy Kwasnikowski Phd., D.Sc., Poznan University of Technology, Poznan, Poland. In this place I would like to thank both for the positive reviews. Professors have wrote, among other things, that topic which is taken into consideration by the author is important and topical. The presented results of the work and the methods used in the book are new and interesting, so that might interest a wide circle of specialists.

Wroclaw, 2015

Contents

Abbreviations and Markings

Q_d	Heat penetrating through walls, ceiling and floor of the cooling chamber
Q_w	Heat dissipated from the refrigerated goods
Q_L	Heat delivered by air, which was introduced into the chamber unintentionally
Q_V	Heat associated with the air cooler fan
Q_{ah}	Any heat while defrosting
Q_{Ma}	Heat generated by lighting, machinery and related equipment in the warehouse
Q_{Me}	Heat generated by people
Q_S	Backup heat due to unforeseen changes of the heat load in the storage
G_i	Load weight (N)
G_k	Weight of carriage and working equipment (N)
G_r	Weight of internal frame (N)
W_t	Work of friction during movement of carriage and mobile frame of the lifting mechanism (N)
l_r, l_k	Distance from the center of gravity from the load with carriage to its gripping and carriage with mobile frame, respectively (l_r, can be assumed equal to l_k)
b_r, b_k	Spacing between retaining rollers of the carriage and mobile frame
d and d_1	Hub diameters of counter-pressure rollers of carriage and frame
D and D_1	Diameters of counter-pressure rollers of carriage and frame
f	Coefficient of friction in hubs
μ	Coefficient of friction of rollers rolling along the tracks
k	Coefficient taking into account other additional resistances (type of wheels and bearings—assumed at 1.1÷1.3)
G_{rg}	Weight of the main frame, including the tilt cylinder

0.75 and 0.25	Coefficients characterizing the approximate position of the frames' center of gravity
$H_{max.}$	Maximum carriage lift height
D_c	Cylinder diameter (cm)
d_t	Piston rod diameter (cm)
p	Pressure (N/cm^2)
I_m	Cylinder capacity (assumed at 0.95)
v_t	Speed of piston movement (cm/min/.)
η_c	Volumetric efficiency of the cylinder, (using rubber and leather seals, approximately equal to 1)
G_I	Cargo weight (N)
v	Carriage lifting speed (platform) (m/s)
η	Coefficient of performance of the device mechanisms
k	Coefficient taking into account the impact of counterweight, equal to 04–06, when none k = 1
ΣW	Driving resistance of stacker crane along the track (N), set as for the track cart
$m \cdot a$	Forces of inertia
$W_t = f_t \cdot Q$	Movement resistances
f_t	Rolling resistance coefficient
F(V)	Drive force
ξ	Rotating mass factor
L	Conveyor length 70
μ	Rolling friction coefficient
v	Speed of cargo movement
η	Transmission performance
l	Length of the transferred load unit
B_r	Shaft length (m)
K = 1 or K = 2	Factors according to type of rollers (normal, heavy)
W_{edi}	Global ratio characterizing the suitability of EDI in organization (J)
w_i	Weight assigned to the i-th partial indicator
W_j	Value of the i-th partial indicator of EDI suitability

List of Figures

List of Tables

Chapter 1
Introduction

1.1 Introduction

In the early 1970s the first modern warehouses were established in the US and Europe. Currently, the world's largest owner, manager and developer, the ProLogis group, owns or manages 2.5 thousand warehouses with a total area of 40 million m^2 (of which 2.1 million m^2 are in Poland). The inclusion of modern warehouses in the logistics chain was to bring down the cost of getting goods to customers, with benefit to all participants in the supply chain. Their innovation was based on the technical standardization, which streamlines the workflow on the one hand, and on the other, the fact that they provide value-added services, which include:

- labeling,
- repacking,
- customs clearance,
- simple production activities, etc.

The process of managing these objects was streamlined by software applications: computer programs [310], and Auto-ID and electronic data interchange systems [148, 151] as well as by technical automation in the field of material flow and storage. This allowed implementing the strategy of supply in the following options:

- KANBAN supply chain,
- Just and Time,
- Within 24 or 48 in door-to-door.

With the economic globalization, warehouses have become increasingly essential. Spot logistics infrastructure consists of four types of facilities:

- ports,
- handling terminals,
- storage centers,
- logistics centers.

P. Zajac, *Evaluation Method of Energy Consumption in Logistic Warehouse Systems*, EcoProduction, DOI 10.1007/978-3-319-22044-4_1

That support a particular type of transport (or type of freight units). Whereas the largest flow of freight units is supported by the two latter types of warehouses. In Western Europe, 50–60 % of the storage space is occupied by logistics companies, 30–40 % by trade hubs (wholesale, retail chains), with production facilities taking 10–20 %. Over time, a "critical mass" of storage space became an important, and is now around 10,000 m^2, so individual warehouses have no reason to exist—hence the observed systematic process of grouping them in one area, into commerce and industry parks, which are further combined to create logistics nodes, and the biggest of those are called logistical hubs, which bring together the activities of companies from many logistics links, modes of transport, research institutes, technology suppliers, etc., as well as a large number of customers. Due to its geographical location, Poland is within the range of a number of hubs, the most important of which are:

- German (Hamburg, Bremerhaven, Frankfurt am Main),
- Dutch (Amsterdam, Rotterdam),
- Belgian (Antwerp),
- Italian (Bologna).

Over the past 18 years, Polish trade interchange with foreign countries has increased in value three times, whereas the turnover (volume) in ports in 2010 remained at the 1995 level. This shows that although the role of ports is significant, it does not grow in the logistics chain. Therefore, by the end of 2007 Europe still continued to observe a boom in warehouses. In the period 2000–07, the warehouse investment market increased 2.5 times, whereas long-term forecasts for Poland and Europe assume a continued growth in demand for logistics services in Europe, which means greater current demand for storage.

Better use of existing resources and the construction of new facilities in accordance with the provisions of European standards and global trends in terms of respect for the ecosystem involves the efficient use of energy allocated to support logistics warehouse processes, which include:

- Production processes,
- Transport of goods,
- Storage.

The concept of energy intensity is connected with the work in the field of zero-energy buildings, hybrid vehicles, thermal modernization, where energy savings are implemented—among others—by:

- use of solar energy to generate electricity and heat,
- efficient office lighting combined with motion sensors,
- efficient lighting in warehouses combined with motion sensors and good daylight exposure,
- increased facade tightness and thermal performance,
- natural ventilation,

- optimizing the daylight exposure in offices, which helps to reduce the consumption of artificial light,
- recovery and re-use of rainwater,
- recovery of heat generated by equipment used in the facility as well as people, and the heat generated by the goods stored in the facility (e.g. fruit),
- the use of devices to limit the consumption of water in toilets [142, 150, 317].

Work is carried out, e.g. [227], to improve energy efficiency and reduce CO_2 emissions in existing terminals, workshops and warehouses. A very important task is to control the implementation of sustainable environmental solutions in all facilities being built or modernized. The energy intensity for warehouses is shown in Fig. 1.1.

For long-distance transport (Fig. 1.2), one may be even use the online calculator [309] to variate the means of transport with regard to CO_2 emissions. There is no similar knowledge and tools for logistics storage systems, no acquisition of statistical data in the economy is carried out on a comparable level. In this regard, there are papers and regulations related to energy efficiency, including e.g.:

- Directive of the European Parliament and of the Council of 5 April 2006 on energy end-use efficiency and energy services and repealing Council Directive 93/76/EEC,
- Renewed Lisbon Strategy,
- National Cohesion Strategy 2007–2013.

The main objective of Directive 2006/32/EC is to achieve an economically reasonable improvement in the end-use efficiency of fuels and energy in the Member States of the European Union through: setting goals, mechanisms and incentives; setting institutional, financial and legal frameworks to remove the existing market barriers affecting the end-use energy efficiency; promotion of programs to improve energy efficiency; development of high-quality energy services for end-users; harmonizing the methodology for calculating and verifying energy savings.

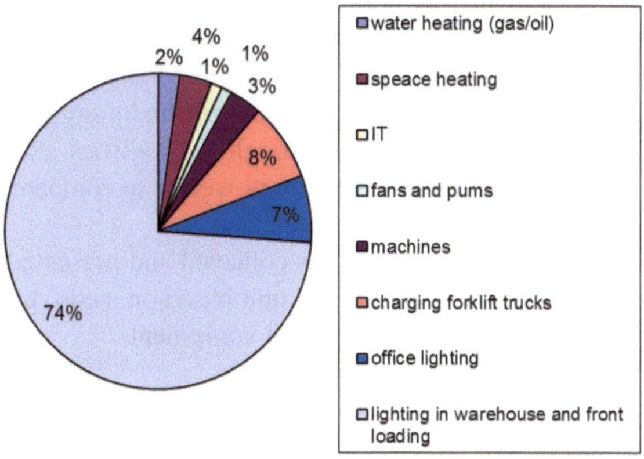

Fig. 1.1 Energy intensity of warehouse processes. Author's own work

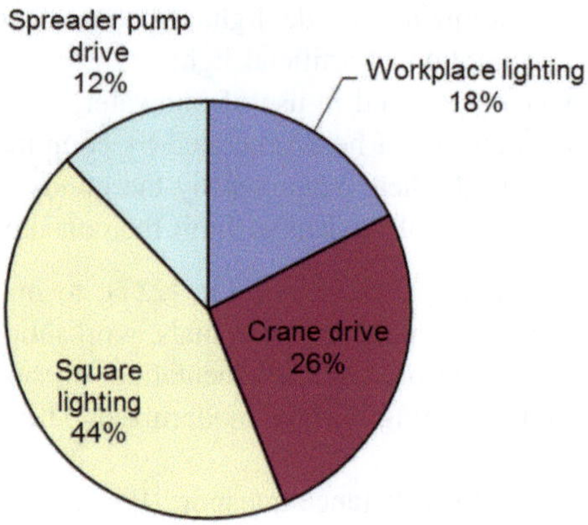

Fig. 1.2 Energy intensity of long-distance transport terminals. Author's own work

The above mentioned Directive obliges Member States to collect and transmit the data necessary for monitoring, evaluation and planning of actions to improve energy efficiency.

Energetics currently use the ODEX index, an aggregate indicator of energy efficiency. It was developed due to the need for monitoring energy efficiency and to achieve a clear, operable and comparable indicator reflecting progress in energy efficiency in the Member States of the European Union. This ratio is obtained by aggregating the changes in energy consumption, observed in a given period of time at certain levels of end use. Using the relevant physical parameters, the ODEX indicator provides a good illustration of progress in the field of energy efficiency.

ODEX is an alternative to monetary indicators of energy intensity that depend on many factors related to—indirectly—energy efficiency. The ODEX ratio does not show the current level of energy intensity, but progress in relation to the base year. ODEX indicators are useful for monitoring the implementation of the indicative target for energy efficiency, as defined in Directive 2006/32/EC.

Figure 1.3 shows the energy consumption of various facilities in the national economy, including warehouses.

In Germany the notion of transport index is also used, which takes into account energy intensity, and whose value is only a hint at how much energy must be supplied in order to obtain the required capacity [145, 319].

The paper then presents a new methodology for evaluating the energy intensity of moving a freight unit through the logistics storage system that allows you to determine the energy intensity as well as to compare the energy intensity of moving a freight unit through a warehouse.

Moreover, the author has collected and presented research results on the energy intensity of moving a freight unit based on a euro pallet through the logistic storage system, with varying technical equipment.

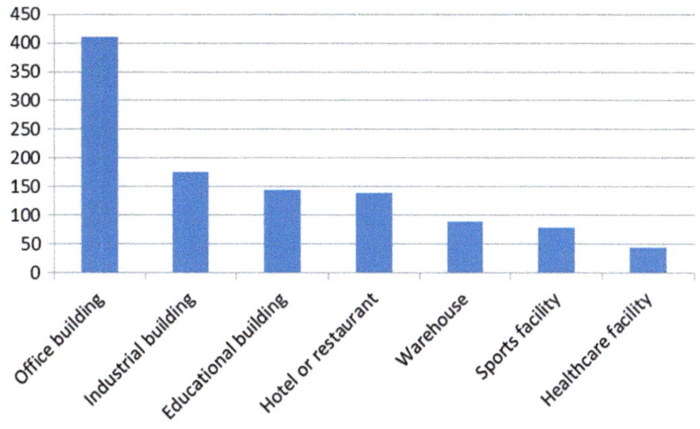

Fig. 1.3 Energy intensity of facilities [254]

Fig. 1.4 Energy intensity of enterprises. Author's own work

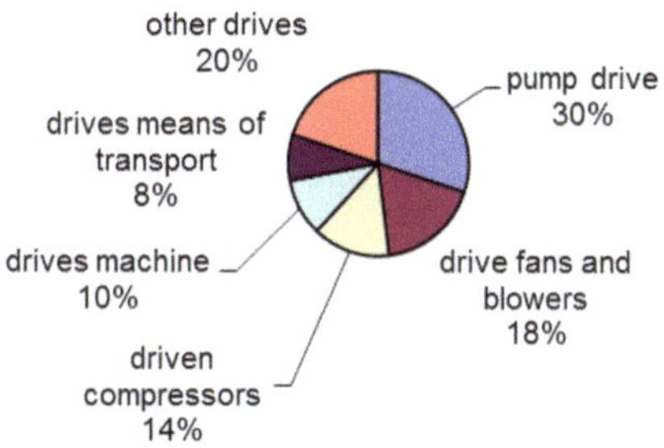

Moreover, back to the issue of CO_2 emissions, the estimated emission of a forklift is 2.4 [T/year], and of a warehouse office—0.08–0.12 [T/year].

In the further section the thesis is presented, as well as a review of literature (state-of-the art), followed by a description of the model and the application of the method to assess the energy intensity using a selected example (Fig. 1.4).

Chapter 2
Thesis

The aim of this study is to develop a model for assessing the energy intensity of the logistics storage system, the storage freight units (pieces), which would allow analytically determining any logistics warehouse management system's (LSM) energy intensity characterizing the handling of a freight unit in the warehouse from entry to exit, which is also sufficient to move the freight units in the processes of acceptance, storage, assembly and release.

LSM energy intensity is energy delivered to the LSM: electricity, gas and fossil fuels, etc. consumed by the loading and information interchange subsystems.

The specific energy intensity of LSM characterizes the determined volume of cargo or annual turnover—wherein the monograph defines contractual modes of work of LSM and the corresponding LSM energy intensity of these processes:

1. The energy intensity of moving a freight unit in the LSM,
2. The energy intensity of the shortest time of handling a freight unit in the LSM,
3. Other, e.g. energy intensity of the freight unit with a maximum LSM gain.

In the LSM, significant impact on the level of energy intensity is delivered by: the mission and strategy of LSM, resulting from the optional division of logistics systems (e.g. military logistics, courier logistics). This issue has been omitted in the monograph for lack of comparative data, but it could be the subject of further research papers.

Theoretical considerations in the paper were verified with the method for the assessment of energy intensity of LSM handling freight units, pallets, pieces, with a module of 800 × 1200 mm, working within the hubs in Poland PROLOGIS and Panattoni Park and the available foreign results.

P. Zajac, *Evaluation Method of Energy Consumption in Logistic Warehouse Systems*, EcoProduction, DOI 10.1007/978-3-319-22044-4_2

Chapter 3
Literature Overview

The subject of energy intensity takes an important place in the world literature and attracts the interest of almost everyone—both scholars in various fields of science as well as companies or individuals. It can be stated that very broad scientific work is done in the field of the energy intensity of technical systems, whose main aim is to reduce the operating costs of these systems.

In addition to working with energy intensity, one can find such papers as [44, 51, 317, 318] on the energy intensity expressed with exergy; a forerunner in this field is Professor J. Szargut [267].

The literature review included work available in the following knowledge bases: EBSCO, COMPENDEX, PROQUEST, Baztech and e-Journals. The monograph's list of references takes into account only those items that are not older than five years and some older items, which the author believes constitute a milestone and are recognized as fundamental in this subject.

To begin with, it is worth noting that the global trends of development in logistic storage systems aim to establish a new lifetime of logistic storage systems (which was previously a period of approx. 25–30 years), which has seen a steady increase in demand for new generations of systems. At the same time, technological advancements lead to the new logistics storage systems most frequently being implemented not based on a significant improvement of existing solutions based on the conclusions arising from operation, but on the basis of a general change in the concept/philosophy of the whole logistic storage system, brought about by new requirements: **ergonomics, ecology, energy preservation, humanization of work, robotics etc**.

Companies managing storage facilities for more than five years are currently revitalizing them. This is impracticable for a single storage, and in the hubs this involves moving from an "old building" to a technically renovated one while maintaining service continuity.

In the environment of the views currently prevailing on the subject, a qualitative evaluation of the operation of products (or to a greater extent technical systems at the design stage) should involve such factors as:

P. Zajac, *Evaluation Method of Energy Consumption in Logistic Warehouse Systems*, EcoProduction, DOI 10.1007/978-3-319-22044-4_3

- Accepted model of assuring the operational quality of the system,
- The investor's promotional strategy adopted for operational qualities of the created system,
- Group classification of the system to be built, resulting from the adopted pragmatics of its operation.

Fitness for operation is mostly a technical issue for the system's designer to achieve the minimum cost of its construction and lowering the technical level so as to not compromise the operating results, or to entail a complaint by an investor. In addition, it is worth highlighting the relevance of the problem purpose affecting the optimization of a logistics system and the satisfaction of the investor, who will expect solutions that minimize costs, while increasing safety and reliability.

A sort of compromise between the technical, economic, social issues (ergology, ecology, safety) involves the generally accepted global technical standards.

Nowadays, the design of logistics storage systems is often based on intuition or years of experience of each manufacture in the creation of a system for the "statistical investor".

The cost of potential behaviors impacts and the operational effects of the system must be predictable at the design stage and taken into account in the implementation stage.

The area where the results of the methods of energy intensity evaluation can used may be managing logistics systems, especially storage, in terms of:

- Allocation of freight units in the warehouse—e.g. the allocation of storage areas, where the currently used method involves the frequency of retrievals,
- Execution of tasks in the assembly process, so far based on the scheduling of these tasks,
- Planning tasks for handling,

and in terms of the revealing design solutions for reloading stations.

In the logistics storage system, one can single out the following elements of infrastructure:

- Fixed (buildings, civil engineering works, ramps, shelves, where they form the supporting structure of the warehouse building, roads, squares, etc.),
- Non-mechanical equipment (shelves, pallets, containers, scales etc.),
- Mechanical equipment (trucks, shelf stacker cranes, conveyors, palletizers, depalletisers etc.),
- Temperature maintenance systems,
- Apparatus and measures for controlling storage processes and performing the processes involving the flow of information,
- Humans,

And four processes (in accordance with [147] et al.) of storage functions, i.e.:

- Admission to the warehouse
- Storage,

- Assembly,
- Retrieval from the warehouse.

An assessment of the energy balance can be found in [80] where the potential energy obtained after loading the container (scrap container type: ACTS) can be recovered during its unloading. This results in a change of the selected powertrain, whose power can be reduced by the potential energy of the container.

3.1 Achievements in the Assessment of Logistics Warehouse and Transport Systems

For the review of the available methods and tools to assess the energy intensity of the logistics storage systems, it is assumed to cover five areas of activity:

- the overall structure of the warehouse,
- the size of the warehouse,
- storage programs,
- mechanical equipment,
- operation strategy.

The overall structure or conceptual design of a warehouse determines the movement of goods in the warehouse, the specifics of individual departments, the flow of information and goods and the relationships between them. Selection of storage system components and their sizing determines the size and dimensions of the storage system and the division of space (not just component) between different areas of the warehouse. The adoption of storage areas, detailed configuration of storage areas, e.g., configuration of movement in the retrieval zone, pallet block—stacking and allocation of vacant space in the reserve storage area, as well as the configuration of automatic storage/retrieval system (AS/RS).

The choice of equipment is determined by the appropriate level of warehouse automation, and identification of the types of equipment for storage, transport, collection of orders and sorting.

The choice of strategy determines how storage—operation of the storage system is handled; for example, in relation to the storage and collection of orders [147]. A case of such strategy may be the choice between random storage or dedicated storage, or collection in the zone, or the choice between sort-while-pick or sort-after-pick. Detailed operational rules, such as batch and path of retrieval, are often not taken into account; these problems are discussed in [94].

It should be emphasized that the design decisions are closely related to the particular warehouse and it is difficult to define a clear boundary between them. Therefore, classification proposals should not be treated as "the one right way".

Operative evaluation is important both at the design stage (in the context of meeting the contracted LSM parameters e.g. energy intensity) as well as warehouse operation. The method for assessing the warehouse in terms of cost, efficiency, energy intensity, use of space and services is an opinion on a particular project or action policy undertaken in conjunction with the requirements—and if necessary, on how they can be improved (in case of regeneration). In addition, the model for assessing the energy intensity of activities can help the designer to make a quick assessment of a number of design options and narrow down their solutions in the design of structures at an early stage. The effectiveness of the assessment method include: a comparative analysis, analytical models and simulation models. This review focuses primarily on the first two: simulation results largely determine the details of implementation. However, this should not obscure the fact that the simulation is still the most widely used technique to evaluate the performance of a warehouse in the scientific literature as well as in practice. Although it is performed at a level far removed from the actual conditions, there are authors who claim that this level is sufficient, e.g. [23].

Research in both directions is very limited; however, examples of solutions and computational tools used to design transport and storage systems and activities will contribute to the elimination of significant differences between academic research and practical solutions, and thus may help to solve the key tasks in the future. Such studies can be found in [94], an addition to previous research studies on storage systems, e.g., [54, 55, 243, 294], etc.

The issue of LSM location (logistics centers/hubs), which is part of a larger area of the logistics system design, is not included in this review of literature. Profile publications on the location methods for storage and transport systems are known, e.g. in [59] et al.

3.1.1 Overall LSM Structure

The overall structure (or conceptual design) of a warehouse defines functional departments, i.e. the number of storage divisions, the technology used and the way of order assembly. At this stage of design, the aim is to meet the requirements for storage and efficiency, minimize costs, which may be discounted value of an investment and future operating costs. One can specify only three published studies devoted to the design of the overall structure.

Park and Webster [207] assuming that the functions are given and provide a selection of types of equipment, rules for storage and order assembly policies aimed at minimizing overheads. The initial cost of the investment and the annual operating cost for each option was estimated using simple analytical equations. Gray et al. [79] raise a similar problem by proposing a multi-stage, hierarchical approach, which uses simple calculations to evaluate and adjust the balance of the design space. Then, a simulation method is used to make a detailed assessment of the performance of individual solutions. Yoon and Sharp [297] propose a structured approach to the use of the design space in order assembly systems, which include such steps as collection of design data, development of alternative design and performance evaluation.

In brief, the published study results on the design of the overall structure of warehouses are limited to the use of a rough estimate or qualitative models combined with limited use of space, which can be reduced by simplifying assumptions. This requires two types of research: (1) assessment based on the complex principles (computer-assisted), taken with incomplete knowledge of the future operating conditions of LSM, and (2) simple, quick-utility models, whose results are less accurate but actually useful in the work of LSM.

It is also worth noting that there is fairly extensive research literature devoted to the issues of the general layout of the facility; see, e.g. Meller and Gau [188].

3.1.2 Capacity and Dimensions of Storage

Storage capacity is essential for the cost of construction, maintenance and replenishment, as well as for handling materials. Previous research has focused on a single storage area and treated decisions on capacity and dimensions as two separate issues.

3.1.3 Storage Capacity in LSM

In capacity modeling, there are two scenarios: (1) Inventory levels are determined externally, so the warehouse has no direct control over the time and the amount of supplies (e.g. an external company warehouse), and any external requirements on storage space must be met by the warehouse; and (2): the warehouse has the ability to exercise direct control over inventory policies (e.g. as an independent wholesale distributor). The major difference is that in the latter case, while solving the capacity calculation problem, one must consider policies and inventory costs.

Assuming that the warehouse has no control over inventory, capacity calculations determine the appropriate size of the warehouse, which can satisfy random demand for storage space.

White and Francis [70, 281, 282] analyzed this issue for one product in a finite planning horizon. The costs taken into account include the cost of constructing a warehouse, storage of products in the warehouse and the storage demand that was not met in the warehouse. The model covers problems with fixed or variable storage space. The second model allows for changes in the size of the storage area in the planning horizon (e.g. by renting additional space), so the decision variables in each period include the size of the storage area. Linear programming is presented for the second model, and optimal solution is identified by solving the problem with network movement (also see Lowe et al. [167]). Similar problems associated with the determination of fixed and variable warehouse size are also discussed by Hung and Fisk [104], as well as Rao and Rao [224], using different cost patterns.

Cormier and Gunn [35, 36], as well as Goh [76] et al. deal with the issue of calculating the size of a warehouse in cases where the warehouse has controlled stocks. Therefore, the costs in their models include not only the cost of building a warehouse, but also the cost of maintenance and replenishment. The paper [161] presented an analytical model for determining the optimal warehouse size for the storage of a single product with a known or random demand. Assuming that there is a possibility to rent additional storage space, Cormier and Gunn propose a closed solution that brings the optimal size of the warehouse, the optimum amount of space for rent in each period, and the optimal size of the replenishments for one product with known demand. Several products are modeled as a linear optimization problem, assuming that the management does not include the times of replenishment deliveries. Cormier and Gunn have developed a non-linear programming model for defining the optimal development of a warehouse in successive periods. Goh et al. have determined the optimal size of a warehouse for one and for many products with known demand. They have studied a more realistic ranged linear model for the cost of building the warehouse in place of the traditional linear cost model. Also, they have taken into account the possibility of joint replenishment in case of storing many products and proposed a heuristic approach to determining the size of the warehouse. The impact of using inventory control policy (e.g. the point of ordering and the size of the contract) on the total required capacity has been presented by Rosenblatt and Roll [235] using simulation.

The ability to answer questions regarding the calculation of the size of a warehouse would have been significantly improved by the use of two types of research. Firstly, the assessment of the requirements for capacity should take into account the seasonality, storage policies and the characteristics of the contract, since these three factors are interrelated, which affects the achievable storage performance, which is the part of warehouse capacity that can be used effectively. Secondly, the capacity calculation models include cost models, and the technical evaluation of these models will be an important input.

3.1.4 Storage Dimensions

Sizing warehouses favors capacity over the surface in order to assess the cost of both construction and operation. The sizing model was first developed by Francis [70], who used the method of continuous approximation of storage areas without regard to the structure of corridors. Bassan et al. [16] extend Francis' concept using the corridors' configuration. Rosenblatt and Roll integrated the optimization model by Bassano et al. with the simulation model that allows estimating the cost of a shortage of storage space, the function of space and the number of zones. They assume simple cycles in order to assess the impact of the size of the store on the operating cost, which can be seen as an attempt to express the energy intensity. This approach is not applicable in warehouses where work with the use of complex cycles is preferred (e.g. restacking, deposition, and release or pick-up of multiple items in a single route).

In this study, attention is drawn to the issue of calculating capacity and sizing, assuming that the warehouse contains a one-product areas. In reality, warehouses may consist of multiple zones/areas, e.g. a forward-reserve complex or different storage sections for different classes of SKUs (Stock Keeping Unit). Different departments should be integrated into one warehouse, they should compete with each other for storage space. Therefore, there are dependencies in determining the overall size of the warehouse and during separation of storage space into departments as well as specifying the size of the warehouse and its departments. Few studies were focused on this kind of dependency. Pliskin and Dori [217] proposed a method to compare alternative allocation of storage space to various departments based on the features of multi-criteria, which clearly capture the relationships between different criteria. Azadivar [7, 8] proposes an approach to achieve the optimum allocation of space between two branches, one of which is efficient in terms of storage, although its efficiency is low, while in the second, the situation is reversed. The aim is to get the best performance of the system by allocating adequate space for these departments to balance storage capacity and dependence in terms of throughput. Heragu et al. [97] consider warehouses with five functional areas, i.e. reloading front at the entrance to the store, release, cross-docking, booking and assembly. They propose a model of optimization and a heuristic algorithm to determine the allocation of SKUs to different storage areas, as well as determine the size of each storage area to minimize the total cost of handling and storage of materials.

The key issue in all studies devoted to sizing is that it requires models for handling performance, which are often independent of the size or layout of the warehouse. Confirming these models requires research or a study of design methods that will explicitly take into account the impact of capacity calculation and sizing on the handling of materials.

3.1.5 Layout of the Area/Storage Areas

This part of the study is focused on the layout of warehouse areas, with particular attention to the storage area. The problems associated with storage are classified as:

(P1) block storage of pallets, i.e. the depth of storage, block height, pallets angle relative to the corridor, space between the pallets, the length and width of corridors,
(P2) layout of the storage department, i.e. the location of doors, orientation of corridors, the length and width of corridors, the number of corridors,
(P3) configuration of the AS/RS, shelf sizing, the number of handling equipment,

The issues of layout affect the performance of the warehouse in terms of:

1. Construction and maintenance costs,
2. Costs of handling materials,

3. Storage capacity, i.e. the ability to distribute inbound shipments,
4. Use of space and
5. Use of equipment.

Each of these issues has been raised in the literature by various authors who take into account the subset of the performance ratios, as presented in Table 3.1.

Table 3.1 A summary of literature on warehouse layout design

Issue	Reference	Aim	Method	Noted
P1	[191]	O4	Analytical models	
	[31]	O2, O4	Analytical models	
	[179, 180]	O3, O4	Simulation models	
	[87]	O4	Heuristic procedures	This applies mainly to determine the depth
	[159]	O2, O4	Heuristic procedures	
P2	[229]	O1, O2	Dynamic programming	
	[28]	O1, O2	Optimum design using shape analysis	
	[239]	O1, O3	Optimal two-dimensional search method	
	[209]	O2	Queuing model	This includes not only the process time but also the waiting time when all units are occupied
P3	[139]	O1, O2 O3	Nonlinear issues, mixed	The model uses the Lagrange equations
	[14]	O1, O2	Nonlinear issues, mixed	
	[237]	O1, O2 O3	Nonlinear issues, mixed	The system is evaluated by means of simulation; if it is not satisfactory, new limits are added, and a solution sought again
	[304]	O1, O5	Heuristic principle	
	[174]	O1, O5	Heuristic principle	Zollinger principle
	[161]	O1	Nonlinear, computer program	To design an automatic carousel system

3.1.5.1 Block Storage of Pallets (P1)

In the case of a problem with block storage of pallets, the primary decision is to choose the depth of storage, in order to balance the relationship between space utilization and ease of storage/recovery, given the limitations associated with stacking SKUs, the size of incoming batches, and products collection schemes. The use of the deep storage method can improve the use of space, as this will reduce the required number of corridors, but on the other hand, it may also result in a reduction in the use of space due to the honeycomb effect, which will create a space that cannot be used for storage until freight units from adjacent tunnels have left the intermediate storage areas, and thus, until the entire line has been depleted. The size of the honeycomb effect depends on the depth of the line and the rate of pick-up of individual products. Therefore, it may be advantageous to store the products of different classes in tunnels of different depth.

To use storage space in the best way, it is necessary to specify and coordinate the depth of storage for various products. In addition to the configuration of the line, the problem of block storage of pallets also affects such decisions as the width and orientation of the corridors, the height of the block and the distance between the blocks. All these decisions, in turn, affect the use of storage space, the performance of handling materials and storage capacity.

The issue of block pallet stacking was raised in several articles. Moder and Thornton [47] examine the ways of stacking pallets in a warehouse, the impact on the use of space and ease of storage and retrieval. They take into account such design factors, as the depth of the storage line, pallet distribution angle relative to the corridor and the distance between storage lines. Berry [48] discusses the relationship between the performance of storage and material handling costs by developing analytical models for evaluating the total storage capacity and the average distance to be traveled for the space requirements for storage. The factors taken into account include the shape of the warehouse, the number, length and orientation of the corridors, the depth of the line, the throughput and the number of SKUs in stock. It should be noted that the models for the total storage capacity and ones for the average distance to be traveled are integrated, and the layout of the storage, which maximizes the warehouse's capacity is different from the system which minimizes the distance to be traveled. Marsh [179] uses simulation to evaluate the effect of alternating lines depths and the rules of allocation of cargo space. Marsh [180] compares the developed layout design using the simulation models designed by Marsh [179] and the analytical models proposed by Berry [22]. Goetschalckx and Ratliff [75] developed an efficient algorithm of dynamic programming in order to maximize the use of space by choosing the depth of the line from a limited number of permitted depths and allocating inbound shipments to lines of various depths. Larson et al. [154] propose a three-stage heuristic approach to the problem of pallet storage layout based on classes, in order to maximize the use of space and minimize the cost of materials handling. The first stage determines the layout of corridors and the size of storage areas; the second stage allocates SKUs to storage configurations; and the third allocates space to storage configurations.

A study on the issue of block pallet storage suggests other rules or algorithms, typically using restrictive assumptions, i.e. assuming that the quantities of replenishments and the retrieval frequency are known for each SKU. In reality, these factors change dynamically; moreover, the SKUs themselves are subject to change, and the block storage schemes optimized for the current conditions may not be optimal in the future. It is necessary to reveal a good solution in case of dynamic uncertainty of storage and retrieval requirements.

3.1.5.2 Layout of the Storage Area (P2)

The issue of the design of the storage area consists in determining the layout of corridors in the storage area in order to minimize the cost of construction and handling materials. Decisions typically include the orientation corridors, their number, length and width, as well as the placement of doors. In order to assess the operational costs, assumptions are usually made about the storage and order picking policies, the most common assumption being random storage and order picking in simple cycles.

While assuming a layout configuration or a small set of alternative configurations, one can create models to optimize each configuration. Roberts and Reed [228] assumed that storage space is available in the form of identical sectors. Bassan et al. [16] consider a warehouse on a rectangular plan, with corridors parallel or perpendicular to the longest wall. Furthermore, they also discuss the optimal placement of doors in the storage area, and the optimal layout where the storage area is divided into different zones. Roll and Rosenblatt expand the concept of Bassano et al. with the additional cost associated with applying a group storage policy. Pandit and Palekar minimize the expected response time to inquiries regarding the storage or retrieval using the queue model to calculate the total response time, including waiting and processing time for different types of layout. Assuming the response times, the optimization model is to set the optimal configuration of storage space.

Roodbergen and Vis [231] present an optimizing approach to the choice of the number and length of corridors and the positioning of the forklift station (stops), so as to minimize the route length during the assembly. They have developed models for both S-shaped routes as well as for the largest gap policy and found that the route policy choice can, in some cases, have a significant impact on the size and arrangement of a department.

The conclusion drawn by Roodbergen and Vis [231] is quite significant, as it also raises the issue of trying to optimize the storage department without knowledge of the actual performance of materials handling. Additional studies are needed that will identify the effect the layout (in the case of departments with the right shapes) has on the total costs during the operation of the facility, taking into account the changing requirements for storage and retrieval.

3.1.5.3 AS/RS Configuration (P3)—Number of Stacker Cranes and Corridors and the Size of Shelves

The problem with the configuration of AS/RS (Automated Storage/Retrieval Systems) is to determine the number of stacker cranes and corridors and the size of shelves in order to minimize the construction, maintenance and operational cost—or to maximize the use of the warehouse equipment. Models of optimal design or practical procedures set out in Table 3.1 usually use the practical expression of the cost based on simple assumptions for operational strategies and a known rates of storage and retrieval.

Karasawa et al. [124] present a non-linear formula of integers, combined with decision variables constituting the number of cranes and the height and length of shelves, as well as costs, including the cost of construction and equipment, while meeting the requirements for handling and storage capacity. Ashayeri et al. [5, 6] solve a problem similar to that described by Karasawa et al. With known requirements for capacity and shelf height, their models can be simplified so as to cover only a single design variable, i.e. the number of corridors. Moreover, the objective function for the number of corridors is convex, which allows the use of a simple, one-dimensional search algorithm to solve the problem. Rosenblatt et al. propose an optimization model, which is a slight modification of the suggestion by Ashayeri et al., which provides support for several corridors using a single stacker (this type of solution is used in the study solutions). A combination is proposed between optimization and simulation, where the optimization model allows creating the initial design, and simulation enables the assessment of performance, i.e. the level of service. If the requirements resulting from the restrictions evaluated through simulation are met, the procedure is stopped. Otherwise, the optimization model is changed by adding new restrictions, constructed by approximating the simulation results. Zollinger [320] proposes a heuristic approach to designing the layout of the AS/RS. Design criteria include the total cost of the equipment, the use of the AS/RS unit, handling time, the number of tasks in the queue, and storage space requirements. The formula in analytical form allows determining these criteria as a function of the number of corridors and the number of levels of the shelf. Malmborg [171–176] uses simulation to clarify the estimates of some parameters that are then substituted into the formulas in analytical form.

While observing the development of LSM in Poland in modernization, investors are increasingly looking for answers to the question whether the above-described solution are competitive to lower storage in pallet racks with AS/RS support. Preliminary analyzes give encouraging results to a detailed consideration, although so far an AS/RS "boom" is yet to be observed.

Designing automated carousel systems was discussed by Lee and Hwang [155, 156, 158, 165]. They applied an optimization approach to determine the optimal number of units to store/retrieve, and the optimal size of the carousel, thereby reducing the initial cost of the investment and the operating costs over a finite planning horizon, the limits of throughput, storage capacity and resulting from the nature of the facility.

There are two major problems that focus further research on the AS/RS:

1. Getting results for a much wider range of technological options, e.g. double-deep racking, multistop stacking cranes with a buffer and 2. The problem of the sensitivity of the AS/RS configuration to changes in the expected period of storage and retrieval rate, or the effects of rapid changes in the product mix.

3.1.5.4 Storage Space Management Strategy

The basic storage strategies include random storage, fixed place method, based on class and frequency of retrieval, as explained by Gu et al. [81]. Hausman et al. [96], Graves et al. [78] and Schwarz et al. [250] compare random storage, fixed-place storage, and based on the class of AS/RS systems using both analytical models and simulations. They prove that dedicated storage can significantly shorten the reloading time compared to random storage, and that class-based storage with relatively few classes allows for shorter handling, which is similar to the results of a dedicated storage. Goetschalckx and Ratliff [74] and Thonemann and Brandeau [269] have theoretically proven that the DOS storage policy is the most promising in terms of minimizing handling costs. Historically, the introduction of the DOS storage policy (Day of Supply) was difficult, because it required tracking and managing each SKU, but modern warehouse management systems (WMS for short) provide this capability. In addition, the performance of DOS storage depends largely on such factors as the variability of demand, balancing the workflow through the warehouse (input-output), inventory control. Kulturel et al. [132] compared class-based storage and DOS using simulation, where class-based storage consistently earned much higher scores. Such a conclusion was possible due to the fact that the assumptions of the DOS model rarely work in practice.

All results relate to the operational strategies of the AS/RS pallet handling system. Other storage systems are rarely studied. Malmborg and Al-Tassan [172] developed analytical models to evaluate the performance of dedicated and random storage in warehouses handling units smaller than the pallet, but have not submitted conclusions comparable to handling pallet units.

It can be concluded that additional studies are required, in particular to clarify the conditions under which the storage policy has or does not have a significant impact on the capacity or the duration of handling.

3.1.5.5 Order Assembly

On any given day or during a given shift, picking multiple orders may be required. These contracts may be similar to one another in many ways, e.g. some are transported by the same carrier or the same type of transport, or have the same date and time of receipt. If there are similarities in the subsets of orders requiring joint shipment, the goods must also be picked up at about the same time to avoid intermediate and temporary storage. Thus, the common practice is to apply wave picking, i.e. releasing a fragment of an order for the day (shift) and expecting that said fragment will be shipped in a specific part of the day (shift).

In addition to wave picking, the widespread use of strategies for shipment picking and zone picking is applied. Shipment picking includes the allocation of contracts to be picked at the same time, in one operation. In the case of the zone picking, storage space is divided into areas, each of which is assigned an employee who performs work in the course of completion only in their designated area. Zone picking can be categorized between sequential and parallel. Sequential picking is similar to the flow line where picking carts that can accommodate one or more orders are transmitted sequentially through the zones, employees in each zone extract the goods within the zone, place them on carts and send them to the next area. Bartholdi et al. [12–14] suggest a chain picking method, which is similar to the sequential method, but does not require the restricting employees to individual area). In the case of parallel picking, orders are carried out simultaneously in all areas. The goods are forwarded to the sorting system and orders are created.

In the picking process, its energy consumption is affected by the following factors expressed in the questions below:

1. Are the goods to be forwarded to the picker ("goods to man"), or does the picker move to where they are stored ("man to goods")?
2. Are orders assembled in waves? If so, how many waves are there and in what period?
3. Is the warehouse divided into areas? If so, will zone picking be sequential or simultaneous?
4. Will contracts be picked in batches, or separately? In the case of batch picking, will sorting take place before or after assembly?

Research on the selection of order picking strategies is rare, which may be due to the complexity of the problem. Lin and Lu [165, 166] compare picking a single order with picking shipments for different types of orders that are classified based on the size of the order and the number of ordered items. Petersen [211] conducted a simulation of five different approaches to order picking: picking a single order, picking a shipment, sequential zone picking, simultaneous zone completion and wave picking. The two control variables in the study with the use of simulation are: the daily number of orders and demand variation, while other factors such as the layout of the warehouse, allocation of storage space and space configuration (in the case of zone and wave picking) are fixed. The performance indicators used to compare the different approaches include: average daily labor, the average length of the day, and the average percentage of delayed orders. For each approach to picking, simple rules are applied for division into batches, determining routes and wavelengths. It should also be noted that performance indicators are mainly related to the order picking efficiency and quality of service; additional costs caused by sorting with the zone and wave picking of the shipment are not included. In addition, the comparison of approaches generally relates to the structure of the contract, while it is assumed that other important factors such as allocation of storage space and detailed implementation of picking policies are fixed. Therefore, the results should not be considered as a generic; more research in this direction is needed in order to provide more guidance for warehouse designers.

The order picking strategy remains a largely unsolved design problem. Additional studies would be appropriate, particularly if they were to allow for characterizing alternative solutions for assembly, which would be easily applied in the decision-making process at the design stage. For instance, would it be possible for researchers to develop performance curves for different order completion strategies?

3.1.5.6 Performance Evaluation

Performance Evaluation provides feedback on the quality of the proposed project or the operational policy and, more importantly, information on how to improve it. There are different approaches to evaluating performance: benchmarking, analytical models and simulations. Benchmarking is the process of systematically evaluating the performance of the warehouse, identifying inefficiencies of the system and suggesting improvements. The Data Envelopment Analysis (DEA) is considered to be a suitable tool for this task, as it provides the possibility of simultaneously listing all relevant inputs (resources) and outputs (results) in order to determine the highest performance limit and to identify the relative shortcomings observed in low capacity warehouses. Schefczyk [249], Hackman et al. [89] and Ross and Droge [237–240] present several approaches and case studies using the DEA method in stock benchmarking. A web-based DEA (iDEAS) for storage was developed by Keck Lab at Georgia Tech University and includes data on more than 200 McGinnis warehouses [186].

Analytical performance models can be divided into two categories: (1) corridor models, which focus on storage systems for individual goods as well as handling and operation times and (2) integrated models that focus on storage systems for multiple goods or criteria, in addition to the time of handling/operation.

3.1.5.7 Corridor Models

Table 3.2 shows the results of tests on models of reloading time for corridor systems. A significant part of the study focused on the expected time of the transshipment for a stacker crane in a type AS/RS warehouse for simple or complex cycles. In both cases, there are studies on three different storage methods: in the case of random storage, SKUs may occupy any place, in the case of dedicated storage, each SKU has a designated location, and in the case of class-based storage, the class of SKUs is assigned a group of storage locations, with random storage allowed in location groups. Complex cycles consist in adjusting deposition and retrieval in order to minimize the number of empty runs for the stacker crane, which may be associated with arranging retrieval and selecting locations for storage. The results in this category typically assume infinite acceleration in order to simplify the

Table 3.2 The results of tests with models of time handling a corridor

	Random storage	Dedicated storage	Storage based on
SKU handling system AS/RS	Single-command Hausman et al. (1976)	Hausman et al. [95]	Hausman et al. [95]
	Bozer and White [28, 29]	Thonemann and Brandeau [266]	Thonemann and Brandeau [266]
	Thonemann and Brandeau [266]	Kim and Seidmann [126]	Rosenblatt and Eynan [230]
	Kim and Seidmann [126]		Eynan and Rosenblatt [62]
	Hwang and Ko [107]		Kouvelis and Papanicolaou [139]
	Lee [154]		Kim and Seidmann [127]
	Hwang and Lee [105]		Pan and Wang [200]
	Chang et al. [37]		Ashayeri et al. [6]
	Chang and Wen [38]		
	Koh et al. [130]		
	Lee et al. [156]		
	Double-command Graves et al. [78]	Graves et al. [78]	Graves et al. [78]
	Bozer and White [28, 29]	Kim and Seidmann [126]	Kouvelis and Papanicolaou [140]
	Kim and Seidmann [126]		Kim and Seidmann [126]
	Hwang and Ko [108]		Pan and Wang [200]
	Lee [154]		Ashayeri et al. [6]
	Han et al. [92]		
	Hwang and Lee [105]		
	Chang et al. [37]		
	Chang and Wen [38]		
	Koh et al. [130]		
	Lee et al. [156]		
	Multi-platform Meller and Mungwattana [189]		
	Potrc et al. [217]		
AS/RS unit with operator em	Hwang and Song [110]		

(continued)

Table 3.2 (continued)

	Random storage	Dedicated storage	Storage based on
AS/RS system at the end of the shelf	Bozer and White [28, 29], Bozer and White [27], Foley and Frazelle [69]	Park et al. [204]	
Revolving and rotating shelves	Han and McGinnis [91] Su [262] Hwang and Ha [106] Hwang et al. [107]		Ha and Hwang [87]
Traditional corridor system korytarzowy	Hall (1993)	Caron et al. [36]	Jarvis and McDowell [116]
	Jawis and McDowell [116]	Caron et al. [35]	Chew and Tang [39]
	Chew and Tang [39]	Jawis and McDowell [116]	Hwang et al. [109]
	Hwang et al. [110]	Chew and Tang [39] Hwang et al. [109]	

travel time models, although more complex models are created, taking into account the acceleration in different axes of movement (see, e.g. Hwang and Lee [106–108]; Hwang et al. [109, 110], Chang and Wen [111], Chang et al. [112]). There are several articles that describe mathematically the issue of deriving travel time distribution (see Foley and Frazelle [68], and Foley et al. [69]). Research on travel time models for carousel systems is consistent with studies on AS/RS.

With the knowledge of the travel time, one can develop models of service time in the AS/RS, taking into account the time required for loading/unloading and placement/retrieval to/from the storage slot. Queuing models were developed, assuming different travel time distributions. See e.g. Lee [156], Chow [39, 40], Hung et al. [104], Bozer and White [27–29], Park et al. [205–209] for an AS/RS system, Chang et al. [37, 38] in the case of traditional corridor systems and shelf-end goods picking systems, see Bozer and White [27], Park et al. In order to assess the capacity of the AS/RS, stochastic optimization models have been developed, with a limited length of storage queue and waiting time after the submission of the request for retrieval (Azadivar [7]).

Throughput in carousel systems was modeled by Park et al. and Meller and Klote [91–93]. Park et al. consider a system comprising two roundabouts and one picker, and derive an analytical expression for the throughput of the system and the picker's workload assuming a deterministic and exponential distribution of picking time. Meller et al. developed capacity models for systems with multiple rotors using a dual server queuing model.

In the case of traditional corridor storage systems (e.g. container racks), two kinds of travel time results were developed: (1) models that provide estimates of expected travel time and (2) pdf models for travel time. These models require assumptions on the structure of the route, e.g. through route (Hall [90]), return route (Han [92] or Caron et al. [35, 36]), or the largest gap (Roodbergen and Vis [231]). While these models are parameterized in terms of the attributes of the storage system design, they can be used to support the design by searching for relevant parameters.

As in the case of AS/RS and rotor, studies were conducted to integrate the travel time models with performance models. Chew and Tang [39] applied their PDF model for travel time to analyze the division of orders into batches and allocation of storage space with a queuing model. Bhaskaran and Malmborg [23] present a stochastic model for evaluating the performance of the handling process in corridor warehouses with approximate breakdown of service time, which depends on the size of the batch and the distance to traverse. De Koster [140] developed queuing models to assess the performance of the warehouse which uses sequential zone picking, and where each bin is assigned to one or more orders and transported by a conveyor. Should it be necessary to extract the container in a particular area, it is transported to the appropriate picking station. After retrieval, the container goes back on the conveyor and is moved to the next station. The proposed network model of queuing allows for assessing such as performance ratios as the throughput of the system, the picker's workload and the average number of containers in the system based on factors such as the speed and length of the conveyor, the number of picking stations and the number of retrievals per station.

An analysis of the throughput of sorting systems was conducted by Johnson and Meller [120, 121]. They had assumed that the introduction is a bottleneck in the sorting process, and therefore affects the sorting system's throughput. This model was later included in a broader model by Meller and Rusell, which integrates picking and sorting in order to balance the relationship the assembly and packaging have with different batch sizes and wavelength. Russell and Meller [243, 244] also show the application of the proposed model in determining whether or not to automate the sorting process and in terms of the sorting system design.

3.1.5.8 Integrated Models

Integrated models combine the analysis of the travel time and the service quality criteria with other performance indicators, such as storage capacity, construction costs and operating costs. Malmborg proposes an integrated model for evaluating the performance of a warehouse in the "pass/book" setup. The proposed model uses information about inventory management, allocation of space in a "pass/book" setup and the division of the warehouse to assess the costs associated with: a deficit of storage space, inventory management, replenishment and shipping, picking orders and in-house replenishment in the shipping area. Malmborg assesses several performance indicators for the AS/RS with two platforms. Malmborg and Al-Tassan [3] present a mathematical model for defining space requirements and the duration of the picking cycle for completion systems of a range including SKUs smaller than the pallet, which uses random storage. The input data of the model are the product parameters, equipment specifications, operating procedures and the setup of the storage area. Malmborg (2003) models the relationship between performance indicators, e.g., the expected total cost of construction and the throughput, using such factors as the size of the fleet, the number of elevators and the configurations of shelving in storage systems using rail transportation.

Analytical models of travel time and performance of storage systems provide an important contribution to the research related to the design of warehouses; there is a wide collection of models. Despite the large number of previously completed results, however, there is no unified approach to modeling the travel time or the efficiency of corridor systems—each system and each set of assumptions leads to a different model. Developing a unified theory of travel time in corridor systems would constitute a major contribution to research.

3.1.5.9 Notes/Comments

There are published industrial case studies that provide not only information about the use of various projects and operational methods in practice, but, more importantly, provide possible challenges for future research from an industrial point of view. Table 3.3 presents these analyses, identifying problems and the types of warehouses that were studied. It is difficult to draw general conclusions based on such a small set of specific cases, although one can conclude that the greatest benefits can be achieved by proper design and operation of the warehouse, see e.g., Zeng et al. [321], van Oudheusden et al. [247] and Dekker et al. [55]. On the other hand, on the basis of these cases, it can be concluded that there are some simple, general rules. For example, the principle of allocation based on the COI ratio (cube-per-order) (Callina and Lynn [122]) omits a number of practical issues, such as variability of weight, the costs of conveyance associated with the product, or relationship between products. Some of these issues have been raised in research (e.g. see Table 3.3 in Sect. 5.2 of Gu et al.), while many others remain unexplode. These cases illustrate the gap between models with limited assumptions presented in the research literature and the complex realities present in most warehouses. There is a need for further analysis of the cases in industry to help the community engaged in research over the issues of storage to better understand the real problems with warehouse design. On the other hand, the results of research carried out on more real data sets will have a more significant impact on practice.

On the market there are many commercial warehouse management systems (WMS), which essentially allow managers to track information about products, orders, space, equipment and employees in the warehouse and provide rules/algorithms allowing the assignment of locations, division of contracts into batches, establishing a picking route, etc. A detailed discussion of these systems goes beyond the concept of this paper.

The successful implementation of research results in the current commercial WMS systems or in design software is a rarity. The prototype systems discussed in

Table 3.3 Summary of studies on case studies

Source	Issue studied	Type of warehouse
Cormier and Kersey [49]	Conceptual design	Warehouse for perishable goods, which requires a "Just-In-Time" strategy
Yoon and Sharp [296]	Conceptual design	Order picking system
Zeng et al. [318]	Allocating storage space, storage sizing, storage procedures and order picking	Distribution center
Kallina and Lynn [121]	Allocating storage space with COI	Distribution center
Brynzer and Johansson [31]	Process flow, division into batches, zone picking	Picking systems feeding materials for assembly lines
Burkard et al. [33]	Determining vehicle routes	AS/RS system, in which the S/R unit may work in any corridor
van Oudheusden et al. [272]	Allocating space, division into batches, determining routes	AS/RS system with an operator in integrated steelworks
Dekker et al. [55]	Procedures for storage and determining routes	Multi-corridor system for manual picking
Luxhoj and Skarpness [168]	Planning the use of staff	Distribution center
Johnson and Lofgren [119]	Simulation by distribution	Distribution center

this section may shed some light on how research results can be used to develop more complex, computer-aided designs for warehouses and service systems.

Perlmann and Bailey [212, 213] have demonstrated design software, which allows the designer to quickly create a set of solutions for conceptual designs, including the shape of the building, the choice of equipment and operational procedures, and choose the best solution based on specific design requirements. To the best of our knowledge, this is the only article devoted to computer-aided design of warehouses.

Several papers were made regarding control systems for warehouse design. Linn and Wysk [165] had developed a specialized AS/RS warehouse control system. The control procedure determines decisions such as allocating storage, decisions on which product you need to retrieve when many products of the same type are stored, how to arrange the storage and retrieval sequence, as well as changing the place of storage. For every decision there are several principles of control and the control policy is created by choosing one rule for each decision in a consistent manner, based on the dynamically changing system of status variables, e.g. the level of demand and the volume of traffic. A similar control system for AS/RS warehouses was proposed by proposed Yih and Wang [278] based on neural networks.

Table 3.4 List of LSM parameters

Warehouse parameters	Description
Owner of the facility	Company "A"
Year of commissioning	–
Class of facility	A or e.g. A/B
Distance from the main thoroughfares	Directly by the motorway
Distance from the centers of large cities	Usually, a few miles from the city center
Ease of access	Good
Total area of the facility (m^2)	50,000
Warehouse space (m^2)	50,000
Warehouse height (m)	Up to 12, existing 8, planned 10
Range of supports (m)	6'12
Storage equipment	Loading docks, fire sprinklers (optional), platform 6t/m^2
Modern docks for trucks	Yes
Railway platform	Optionally, project agreed
Fiber optic	Yes
Telephone lines	Yes
ISDN	Yes
Water connection for production	Depending on needs, yes
Sewage treatment plant	Municipal sewer: rainwater, sanitary
Gas connection	Yes
Connection of electric energy (MW)	Yes, depending on needs: possible to supply energy from PZO
Heating	Gas panels installed under the roof
Fire sprinklers	Optional (yes)
Ability to conduct light manufacturing	Optional
Cooled storage facilities (cooling, m^2)	To be agreed
Storage facilities for the storage of hazardous materials (m^2)	To be agreed
Office space (m^2)	15,000
Number of parking spaces for trucks	To be agreed
Number of parking spaces for cars	To be agreed
The smallest unit available for rent (No.)	400 warehouse 30 office
Real estate agents renting warehouse space	"Rental" Agency
Additional services provided by the facility's administrator	Full management and facility administration
Planned area (m^2)	25,000
Expandable in accordance with the wishes of customers	Up to 50,000
Additional information	Warehouse is located in the economic zone

Ito et al. [114] propose a system based on intelligent agents for warehouse modeling, consisting of three subsystems, i.e. a communications system based on agent technology, a reloading system based on agent technology, and a planning and inventory control system based on agent technology. The proposed system based on agent technology is used to design and implement of warehouse simulation models. Kim et al. [127, 128] present a system based on agent technology used to control the storage of cosmetics. In addition to providing communication functions, groups of agents make decisions regarding the operation of their assigned warehouse units dynamically and in real time.

The lack of research prototypes for computer-aided warehouse design is puzzling, given the rapid development of computer hardware and software. Computer-aided design in other areas is enjoying continual popularity among researchers, particularly when it comes to the creation of computational models to support decision-making. On the other hand, there is still a demand for this in the field of warehouse design.

The clearly conclusion is that the studies concerning the design of warehouses focus on analysis and not synthesis, with particular regard to storage systems. This is surprising, although more surprising is the conclusion that only 10 % of the articles directly raising the issue of decisions in warehouses design were published after the year 2000. Given the rapid development of computer hardware and solutions to optimize, simulate and solve mathematical problems, one would expect a broader literature on design.

In literature [1–6, 8–10, 12, 13, 16–18, 43, 58, 62, 67, 68, 70, 72–74, 76, 84, 89, 92, 106–108, 111–113, 124–128, 150, 164, 166, 167, 178, 188–190, 193, 196, 198, 199, 201, 203–207, 227, 228, 230, 231, 247, 253–256, 258, 259, 264–266, 268–270, 272, 273, 280–282], there are a number of parameters that are most commonly mentioned in the design and operation of LSM. A list of those is given in Table 3.4.

The wide variation in the material flow processes, a variety of sets of activities in the functional areas of the store and the possibility of their technical implementation are forcing designers/logisticians of transport and storage processes to accurately analyze and optimally select the assumptions regarding the functions performed by each area of the storage system.

Chapter 4
Methods to Assess the Energy Consumption of LSM

The beginning of the logistics storage system is its mission statement, which is reflected in the strategies—the path of its implementation, including a logistics strategy (Fig. 4.1). After that, a technical design is made based on the processes that undergo computer-aided integration (Fig. 4.5). The new LSM solution is implemented in a space prepared for this purpose—the storage structure (see Fig. 4.2).

Warehouses broken down by classification:

A warehouse can be regarded as the available space, filled with loading units packed on carriers sized 800 × 1200 mm and with a height of 2000 mm in Fig. 4.3 (Fig. 4.4).

In practice, the most common two storage technologies are:

- Simple storage, characterized by block storage, racking equipment and forklifts,
- Automated: high storage, automatic conveyors, stacker cranes.

While analyzing these complex technical systems, it can be concluded that they implement a specific technical process in which energy, materials and information is being transmitted or processed, and their changes are taken into account over time. Only then it makes sense to analyze energy, material, and information (Fig. 4.5).

Activities performed in a warehouse:

- Storage,
- Transportation,
- Packaging,
- Cross-docking.

All devices performing warehouse processes create a logistics system [200]—their use is considered in a multifaceted manner, with respect to technical conditions. The complex relationship is shown in Fig. 4.6.

The energy intensity of LSM is defined as the ratio of energy consumption in the process of handling a freight unit in LSM (passing through LSM) in respect to the number of freight units, in which the energy is involved, or otherwise as the relationship between input and its effects. This statement is equivalent to the ratio of energy put into specific manufacturing or operational activities etc. to the effect observed. Referring the energy intensity to a single freight unit, we are looking

P. Zajac, *Evaluation Method of Energy Consumption in Logistic Warehouse Systems*, EcoProduction, DOI 10.1007/978-3-319-22044-4_4

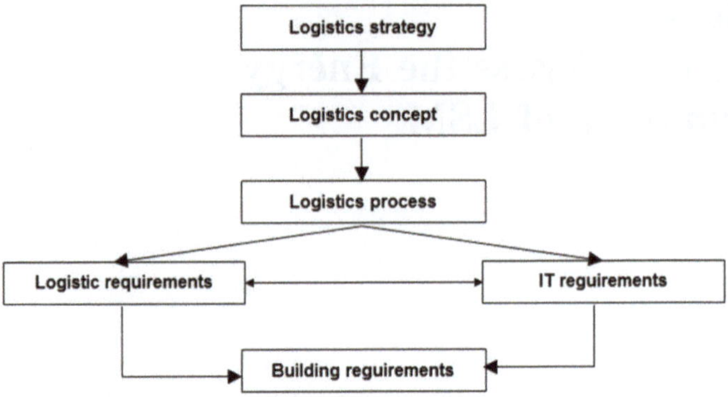

Fig. 4.1 Steps involved in creating LSM. Author's own work

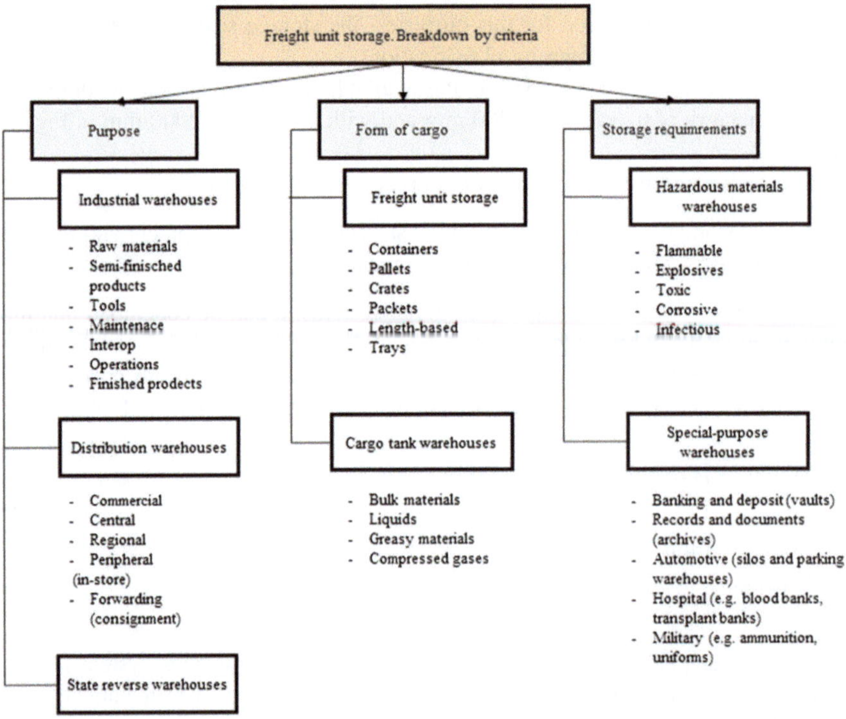

Fig. 4.2 The basic division warehouses based on [139]

trying to answer the question: how much energy is needed to handle it in the LSM? In LSM it is the energy supplied to the handling equipment and lighting, heating, cooling, automatic identification, EDI. Energy delivered to the LSM is not only consumed, but with appropriate technological solutions, is also recovered and stored to be used again for useful work.

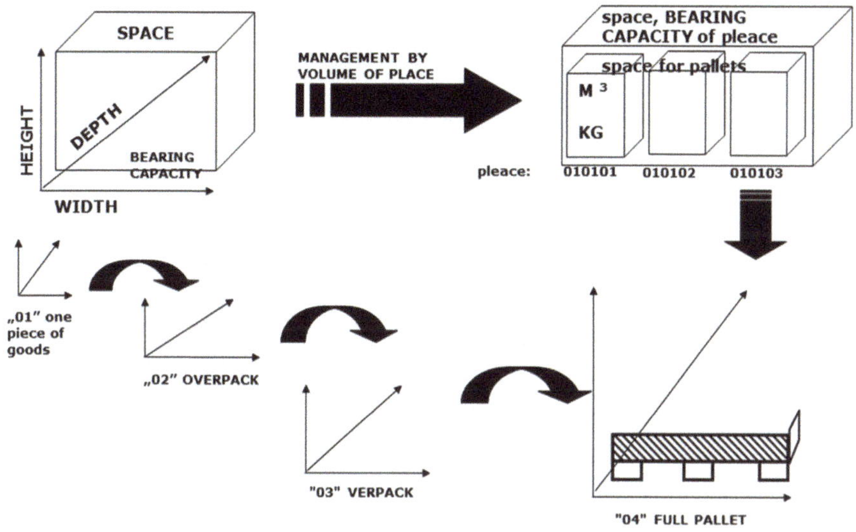

Fig. 4.3 From a single unit to warehouse space—the principle of filling up LSM space

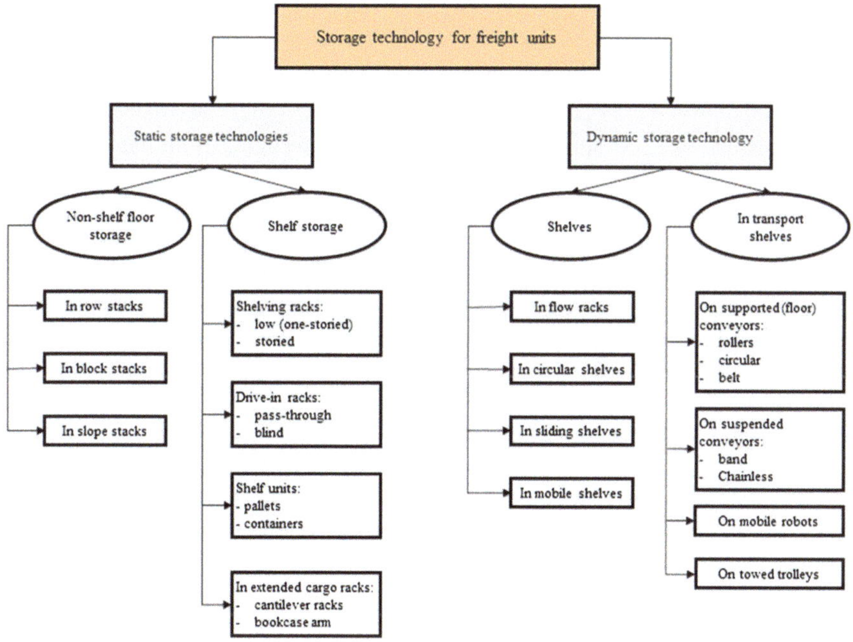

Fig. 4.4 Storage technologies in LSM [135]. Author's own work

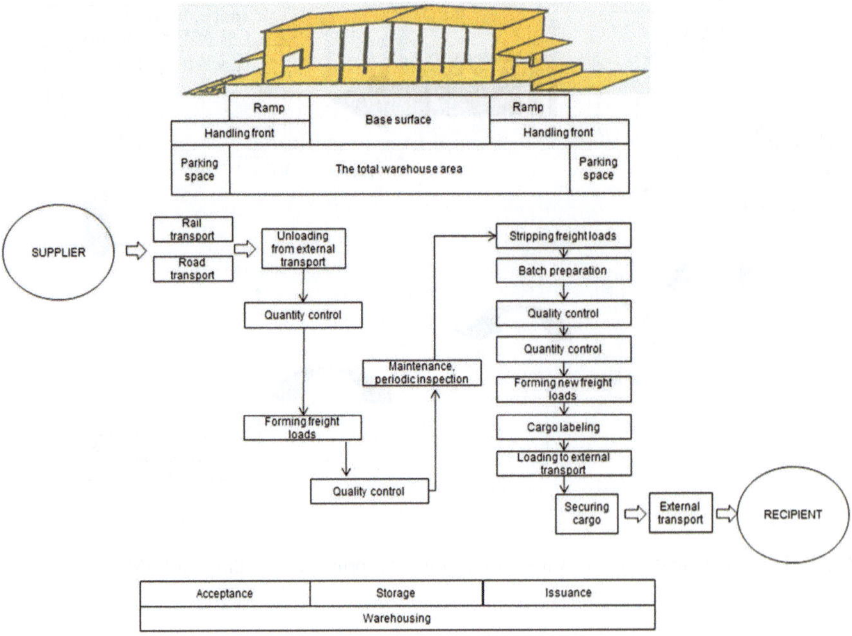

Fig. 4.5 LSM draft, taking into account the system's characteristic features. Author's own work based on [135]

In the work of LSM subsystems involves losses resulting from energy conversion efficiency, as shown in Table 4.1.

LSM often uses combustion-powered equipment (e.g. forklifts) (efficiency 0.36) or electric equipment (efficiency 0.92). LSM uses the energy of fossil fuels, which include: leaded petrol, diesel, LPG, CNG, gas, geothermal and solar energy, fuel cells.

Leaded petrol, diesel and gas are most commonly used to power the materials handling equipment and machines. There are currently research prototypes of forklifts powered by hydrogen cells—as presented at international fairs (e.g. Hanover, May 2012) (Fig. 4.7).

Geothermal and solar energy is only used occasionally as a power source in the LSM, due to high installation costs and little interest in these technologies from investors in Poland. This process is hampered by the lack of legislation, or incentives to its application. However, it is becoming more widespread e.g. in Germany.

Apart from fossil fuels, LSM is supplied with electricity to power transport and auxiliary equipment [313, 316 and others].

Energy intensity significantly affects the costs of LSM operation. Analyzing them, one can divide them into the cost of land and buildings, the cost of personnel and the warehouse equipment costs. For simple a warehouse, the cost would

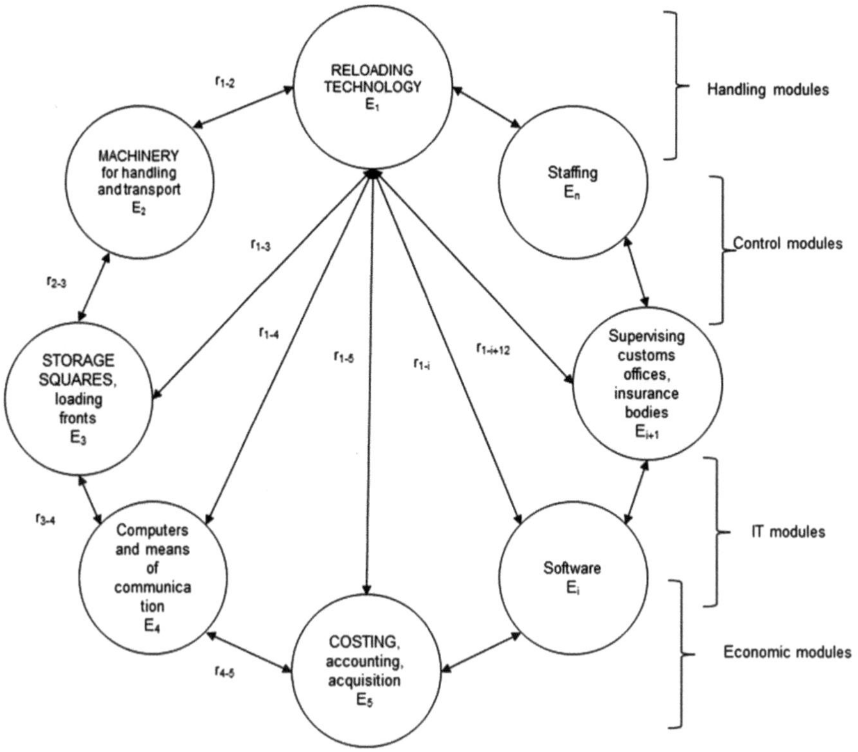

Fig. 4.6 Logistics storage system—components [136]

involve the cost of personnel, then land and buildings, the cost of equipment being the lowest (Fig. 4.8).

In the second case (b) the cost of equipment is very high, the cost of personnel very small, while the cost of land and building is slightly greater in relation to the variant in figure (a). There is a substitution relationship here, referring to the relationship between the three elements. Accordingly, balanced LSM solutions (c) are sought.

Parameters of individual storage areas must be defined and precisely calculated. The technologies used in the facility for internal transport and storage should correspond to the specifics of the handled goods (data contained in warehouse technological design, logistics design prepared by an interdisciplinary team of specialists). Experience shows, however, that even a warehouse built based on the previously prepared technological design may at some point start to fail. Corridors between racks become piled up with freight units awaiting placement in shelves, employees work overtime. The quality of implementation of warehouse processes often decreases as well: the goods are often damaged, there are an increasing number of errors in the order picking… In extreme cases, the entire supply chain

Table 4.1 The efficiency of some conversions [80]

Energy conversion	Device	Efficiency
Mechanical energy	Water turbine	0.90
	Wind turbine	0.46
Heat → mechanical energy	Steam turbine	0.40
	Steam engine	0.20
Fuel → mechanical energy	Spontaneous combustion engine	0.36 and
	Spark-ignition engine	0.25 and
	Gas turbine	0.30
Fuel → » heat → mechanical → electricity	Steam power plant	0.40 and
	Steam-gas system	0.55 and
	MHD generator	0.60
Mechanical energy → electricity	Generators	0.99
Electricity → mechanical energy	Electric engine	0.92
Electricity → heat	Heater	1.00
Chemical energy →	Battery	0.70
	Fuel cell	0.60
Fuel—heat	Steam boiler	0.88
Solar energy → electricity	Solar cell/	0.12

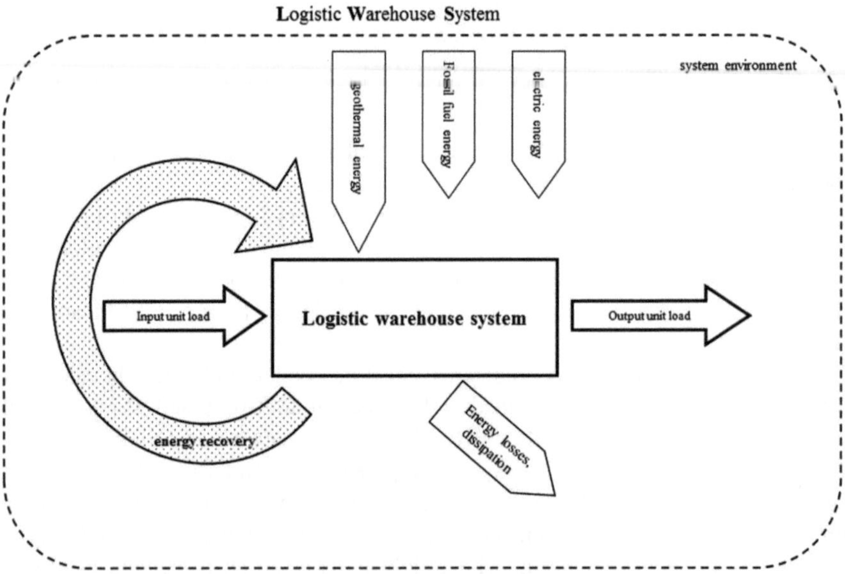

Fig. 4.7 Diversification of energy supplied to the LSM. Author's own work

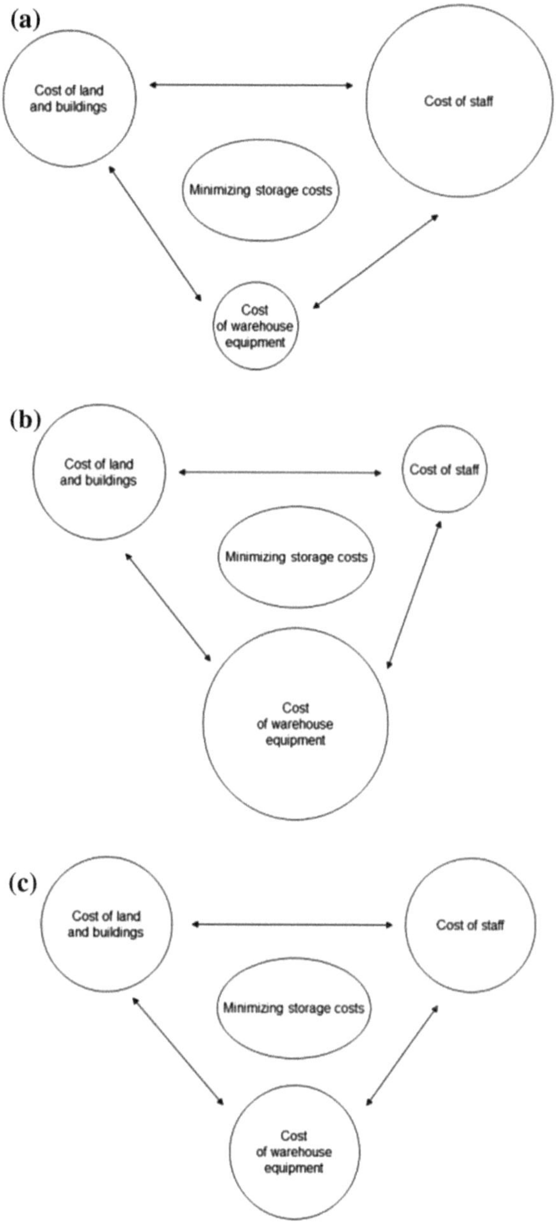

Fig. 4.8 Storage costs for warehouses: **a** simple, **b** automated, **c** balanced. Author's own work

becomes ineffective, which exposes the company to financial losses and the loss of customer confidence.

The unique nature of each LSM requires prior identification of the specific situation of the company, determining the trends and forecasts of development, collecting and analyzing a large variety of figures [228]. Sometimes, however, despite the best efforts on the design and implementation of the investment, a previously efficient warehouse may begin to function poorly. In such case, it is difficult to say whether it is the fault of the investor, the designer or the warehouse employees. After all, it is natural that over many years of operation, the conditions in which the warehouse operates would change. The number of goods the warehouse handles every day is growing, customer expectations are rising. The employee qualifications are decreasing, or the parameters of freight units are modified. Reasons can be many.

In this situation, the only chance to maintain market competitiveness of the company is to modify:

- Processes. Research revealed a common phenomenon, where employees carry out their assigned tasks in a manner they find convenient. However, such a seemingly optimal realization of the process at one workplace may hinder further operations in the logistics chain. A Case is the preparation of goods for shipment, where after inspection (qualitative and quantitative) comes packaging (security for transport). The inspection usually requires breaching the packaging, which in case of hasty and sloppy performance, may significantly impede preparations for distribution. The right approach here seems analyze the LSM chain system and then modify the individual processes.
- Use of organizational and technical measures. SKUs piling up in the corridors between racks… this image is commonplace in many warehouses. This phenomenon does not necessarily result from infrastructural deficiencies or lack of technical measures. The source of this phenomenon may be insufficient number of warehouse workers. The simple shortage of labor can result in staff "saving time" by arranging freight units at random while many shelving slots remain unfilled. As a result, searching and preparing the goods for distribution takes more time than it should, and the proverbial loop closes. In this way, the savings made were merely an illusion. As experience shows, an important part of improving storage performance also involves a systematic upgrading of skills. After all, knowledge and skills ensure the effective use of the available organizational and technical measures.
- Technology. A commonly observed phenomenon is the technology incompatibility between handling/storage and the current needs. One Case could be the insufficient capacity of the forklift truck or the size of the fork. This situation usually results in longer transportation cycles, and often threatens the health and lives of employees. Depending on the needs identified in the transport and storage as well as on the infrastructure available, this can be remedied by replacing some equipment or modifying their parameters (e.g. replace rack beams, replace forks, replacing laser scanners with radio terminals).
- Facility parameters. Given the size of the necessary investment, this method of improving warehousing logistics should be considered a last resort. It is worth to conduct analyses in the direction of increasing the productivity of facilities previously operated by streamlining processes or modifying the parameters of technical measures. However, when this fails to secure a significant increase in

storage efficiency, one should not start the investment immediately. It is worth carrying out further analyses, simulations and draw up a process design for the new facility.

In such cases, one can use the model for assessing the energy intensity of LSM. In the model for evaluating the energy consumption, LSM treat every freight unit individually, tracking its route from the moment of acceptance at the reloading front, through subsequent movements in the course of warehousing, to retrieval at the reloading terminal at the output of the LSM.

Based on the published ([147] and others) division of warehouse processes taking place in appropriate storage areas (zone of delivery, acceptance, storage, picking, release, expedition) into processes:

- Loading (loading, reloading and unloading),
- Physical flow of goods (movement, transport, transit),
- Storage (acceptance, storage, picking, release and expedition),

and based on experience, it is assumed that when it comes to operating equipment in the various areas of LSM, these changes took place several times in terms of technique and technology. Therefore, the computational model takes into account the need for changing handling equipment.

The model includes two changes of transport means for processes handling inputs and two outputs, with the ability to adapt to any given case.

Energy not fully used for work, resulting from the performance of individual components of the system can be recovered in the LSM or converted into another form—further exploited, contributing positively to the reduction of energy intensity.

The energy intensity of technical systems is affected by the information interchange subsystem:

- Automatic Identification System, e.g. barcodes, RFID tags,
- EDI information interchange system,
- Computer-aided systems for warehouse management.

These issues have been characterized in generic and functional terms in [152, 157] and the computational model included them in the form of indicators.

In the case of static storage for postponement of the freight unit, its energy balance does not change. In dynamic storage, which is when the freight unit after deposition can move within the rack or with the rack, the unit's energy changes accordingly.

When using devices such with energy recovery e.g. forklifts or other equipment, the energy balance changes accordingly.

Using the evaluation model of energy intensity must be combined with following the rules found in the literature [to be supplemented] and taking into account the logistics strategies, which the author mentioned in the book [152].

The model allows us to analyze the LSM equipped with the most commonly used materials handling equipment such as forklifts, conveyors and shelving.

Forklifts can be of any type, with any drive. Computation for working forklifts also takes into account their degree of load.

The model does not take into account the energy intensity of:

- Building lighting,
- Energy receivers in the office (e.g. a kettle),
- Insulation performance (e.g. the degree of warming in warehouse buildings, the use of control devices in heat sources such as radiators,
- The facility's sun exposure, etc.

The energy intensity of LSM is estimated from the process of accepting an SKU at the warehouse, through the processes of storage, picking and release. The freight unit at the input terminal has the energy consumption equal to zero. During its further movement through the handling equipment horizontally and vertically, a change in energy intensity values is registered. The efficient implementation of movements is provided by handling equipment which carry out these processes using electricity or/and fossil fuel energy. Lifting a freight unit from a lower level to a higher one requires feeding energy to the technical system implementing this process. The reverse action requires supplying energy to the device—e.g. a forklift carriage, and after retrieval, it only requires enough energy to sustain the unit's lowering speed at max. 0.2 m/s. Energy intensity in horizontal movement is calculated using the equations of motion taking into account the current status of equipment or machinery—loaded or empty. This is shown in Fig. 4.8.

4.1 Description of Model Evaluation of Energy Consumption LSM

Based on the discussion in the previous sections of this paper and e.g. on [12,13, 139], a logistics system can be semantically represented as:

$$SL = SL\ (Z,\ P,\ M,\ T,\ D,\ R) \tag{4.1}$$

where:
Z—supply,
P—production,
M—warehouses,
T—transport,
D—distribution,
R—relationships between the system's components.

Supply (Z), e.g. based on [12], can be expressed as:

$$Z = Z(BI, PM(PZ, PN), ZZ, OD) \quad (4.2)$$

where:

- BI—information database: forecasts, programs and plans for the sale of products and their components; technical documentation, including individual and collective standards (indicators) for the consumption and materiel, lists of typical parts (i.e. commercial) and special parts (purchased under office cooperative), lists of recommended ranges of materials to use…; categories of materials available on the market, price lists, information, offers, advertising brochures, all the information from exhibitions and fairs; lists of suppliers (names, addresses, phone and fax numbers…) with any information on pricing, discounts, lead times, reliability, quality of products,…, material indexes, lists of stations and positions that receive individual materials;
- PM—material needs,
- PZ—dependent needs (demand for raw materials, materials, components, assemblies, …, resulting from the demand for another item processed in the company),
- PN—independent needs (external demand (market demand));
- ZZ—sources of purchase, minimizing the costs associated with the purchase of materials and maintaining inventories as well as creating the basis for the smooth production (unit price of products, delivery dates and their flexibility, and specifying accuracy (day, week, decade, month); distance from the supplier, the quality of materials (compliance with the agreed standards or technical conditions), the terms of payment);
- OD—organization of supply (delivery schedules, means and conditions of transport, type of packaging, supply regulation, quantitative and qualitative acceptance of supplies, financial settlements related to supplies).

Production (P) can be interpreted as follows:

$$P = P(TP(M, S(WS, SS, MS), J), OP) \quad (4.3)$$

where:

- TP—type of production (the scale and degree of specialization of production cells (up to and including workstations), taken together):
- M—mass production is to produce at a certain period the same products in a narrow range, the production volume being well-adjusted to the constant and high demand, or to produce several products (e.g. canned fish with different raw material by a single or very similar technology). A specific variant of mass production is continuous production, e.g. the manufacture of wire, cables,

- S—serial production consists of periodic production of a specified number (series) of similar products using similar methods and means of production (e.g. the processes of casting, mechanical engineering):
- WS—large series production is 1–10 operations on individual production stations,
- SS—medium series is 10–20 operations on production stations,
- MS—macro series is more than 20 operations on production stations and short intervals;
- J—discrete production involves the manufacture of single copies of products of different design, size, equipment, tools, service life (unique products, very complex products having, and those of high value);
- OP—the manner of the flow of parts, assemblies and products between workstations in the production cycle.

A warehouse (M) within the meaning of [137] is an organizational and functional unit, intended for the storage of inventory; it occupies a separate space equipped with appropriate technical measures, managed and operated by a team of people; and according to [2], it is a structure designed and constructed in such a way as to fully protect the stored goods against the loss of quality and quantitative loss, and to ensure safe working conditions for warehouse personnel. Also, the author of [2] distinguishes the following types of storage:

$$M = M(OT,\ PT,\ ZJ,\ SP) \tag{4.4}$$

where:
- OT—open or enclosed squares with dirt or paved surface,
- PT—semi-open, i.e. buildings partially separated from the surface with a protective partitions, which include: basements, attics, roofs, sheds,
- ZT—closed to fully secure the raw materials from the harmful effects of the weather, or other damage to property; generally provide a suitable temperature, humidity and ventilation,
- SP—special, i.e. warehouses adapted for storing only one type of material, or built according to a special design and equipped with special devices; these include: granaries, silos for bulk materials (e.g. cement), cold stores, bunkers.

Warehouse equipment (WM) within the meaning of [137] can be expressed in the form:

$$WM = WM(US,\ STM,\ PUM) \tag{4.5}$$

where:
- US—storage equipment (shelving, racks, hangers, special equipment)
- STM—storage transportation (trucks, stacker cranes, cranes, hoists, gantries, conveyors, manipulators),

- PUM—auxiliary warehouse equipment (for changing the load; for mechanically forming and securing cargo, facilitating manual operation of racks; used in reloading fronts, to identify materials, for measuring the weight, quantity and size, for management and communication).

Transport (T), based on [17, 117, 118], may be systematically understood as follows:

$$T = T(TS, TK, TM, TWS, TL, TP, TKO, TB, TMD) \tag{4.6}$$

where:
TS—road transport,
TK—rail transport,
TM—maritime transport,
TWS—inland waterway transport,
TL—air transport,
TP—transmission transportation,
TKO—combined transport,
TB—bimodal transport,
TMD—multimodal transport.

Road transport (TS) may be presented in the form:

$$TS = TS(f, STS, INL, INP, TP, O) \tag{4.7}$$

where:
I—essential features of road transport:

- best availability of space resulting from the largest network density and cohesion of all modes of transport,
- best adaptation of the road network to deploy demand and supply markets,
- very favorable offer in terms of travel time, resulting from a relatively high operating speed and the best availability in time,
- specialized rolling stock designed to carry loads of various transportability,
- best possibility of commutation to carriers in other modes of transport,
- relatively high cost of movement resulting from a weak degression in unit costs;

STS—means of road transport:

- tractor units, trucks;
- machinery and equipment in cargo works (excavators, loaders, gantries, cranes, truck cranes, conveyors, loading vehicles, transport tippers and lifters).

INL—linear infrastructure:

- national, provincial, local urban, municipal, and in-house roads;
- public roads, express roads and highways.

INP—nodal infrastructure:

- passengers: coach stations, railway stations and bus stops;
- cargo: public discharge areas, stations, squares, and transshipment points;
- means of car transport: technical stations and material—technical resupply stations for cars.

TP—freight transportation technologies:

- Unified;
- Specialized.

O—type of fees for:

- shipping services;
- transport work;
- hire;
- cargo services.

Rail transport (TK) can be expressed as follows:

$$TK = TK(I, KST, INL, IN\ P, TP(PMO, PMN, PKM, PN\ N), FPL, O) \quad (4.8)$$

where:
I—essential features of rail transport:

- mass transit capability,
- relatively low freight rates for medium and long distances due to a strong decrease in unit costs,
- relatively extensive rail network, well suited to the location of the major demand and supply markets,
- high availability of space resulting from a significant density of the road network and transport nodes,
- advantageous offer in terms of travel time, resulting the high reliability of the rail and the frequency and regularity of the connections offered,
- specialized rolling stock suitable for freight transport with different transportability,
- access to carriers in other modes of transport,
- relatively lower security in case of cargo sensitive to shock and reloading, and high risk of theft;

KST—railway transport measures:

- sheeted, open, tanks, jar wagons, special, technical and commercial,
- tractive vehicles (locomotives, diesel locomotives and motor cars, electric locomotives and electric sets);

INL—linear infrastructure:

- railway (electric, steam and petrol traction),
- tanks with loading and refueling points,
- engineering structures,
- rail traffic security equipment,
- equipment for storing information about rail traffic;

INP—nodal infrastructure: stations, cargo points (regions), railway sidings:

- cargo: transportation, handling, storage, warehousing, sorting, grouping;

TP—freight technology;
PMO—mass transport to mass customers (specialized transportation of coal, building materials, ores, etc.; common use of specialist warehouse);
PMN—mass transport to recipients in small batches of cargo carried by block train in shuttle service;
PKM—multimodal and container transport carried out by special wagons equipped with devices for securing the container or semitrailers;
PMN—non-bulk freight, unusual cargo, requiring trains of different types of warehouses in marshalling yards;
FPL—forms of cargo in railway transport:

- regular,
- shuttle,
- block trains;

O—determining the type of fee:

- workforce metrics,
- release workforce metrics,
- energy metrics,
- devices for timing and maneuvering, loading, unloading, carriage,
- railway traffic management, depots.

Maritime transport (TM) is described as follows:

$$TM = TM(I, STM(U, W, P), INL, INP, TPŁ(D, M), O) \tag{4.9}$$

where:
I—essential features of maritime transport:

- ability to carry bulk cargo with the widest range of transportability,
- global reach of transport routes,
- best price of transport over long distances which is the result of the strongest degression in unit costs,

- the small operating speed of ships and the relatively low frequency combined with the timeliness of maritime connections,
- relatively low safety of cargo sensitive to moisture, longer delivery times and transshipment,
- the need to use pick-up/transport services due to the relatively low spatial availability of seaports;

STM—sea vessel:

- U—universal,
- W—specialized,
- P—shuttles;

INL—linear infrastructure:

- waterways;

INP —nodal infrastructure:

- port aquarium,
- port territory,
- port network of railways, roads and port stations,
- port networks;

TPŁ—cargo transport technology:

- D—groupage,
- M—mass;

O—determining the type of fees:

- freight and traffic management,
- fuel, oil, grease, water,
- port and commercial,
- special cruise fees,
- reloading,
- storage of goods in port warehouses,
- handling and in-house transportation,
- towage, pilotage, mooring,
- various other services regarding ships.

Inland water transport (TWS) can be described as follows:

$$\text{TWS} = \text{TWS}(/,\ \text{STWS},\ \text{INL},\ \text{INP},\ \text{TPŁ}(\text{D},\ \text{M}),\ \text{O}) \tag{4.10}$$

where:
I—essential features of inland water transport:

- ability to carry bulk cargo of low value,

- low freight rates due to the large degression in unit costs while transporting large consignments over large and medium distances,
- long delivery time, which is the result of small operating speed and irregular traffic as a result of the dependence on weather conditions and climate,
- poor availability of space associated with the inadequacy of the waterways network to deploy the demand and supply markets,
- relatively low safety of cargo sensitive to moisture, longer delivery times or transshipment;

STWS—inland water transport measure:

- river ships,
- self-propelled barges,
- pushed systems,
- “combi” systems (motor boat with a barge without the drive);

INL—linear infrastructure:

- system of waterways,
- shipping channels;

INP—nodal infrastructure:

- river ports (aquarium, coupe, docks, outer harbor), territory (quays, piers, storage yards, transport ways),
- reloading (handling equipment);

TPŁ—cargo transport technology;
D—groupage;
M—mass;
O—determining the type of fees:

- propellants,
- navigation (lockage, anchorage, channel, pilot),
- port,
- fleet rental.

Air transport (TL) can be expressed as follows

$$TL = TL(I,\ STL,\ INL,\ IN\ P,\ TPŁ, O) \tag{4.11}$$

where:
- essential features of air transport:
- ability to carry relatively small batches of products of a specific natural, technical and economic vulnerability,
- most advantageous time offer, especially on long routes, resulting from the highest operating speed offered and the large and ever-increasing degree of speed during air transport as well as its reliability,

- very high cargo safety, especially of products sensitive to prolonged transport, moisture or vibration,
- the need to use pick-up/transport services due to the relatively low spatial availability of airports,
- high cost of movement at a very large degression in unit costs over long distances;

STL—means of air transport:

- airplane;

INL—spatial linear infrastructure:

- airways,
- control areas of airports;

INP—nodal infrastructure:

- landings,
- airstrips,
- airports,
- international airports;

TPŁ—cargo transport technology:

- complementing aircraft capacity;

O—determining the type of fees:

- takeoffs and landings,
- external services,
- special services,
- maintenance,
- commercial.

Transmission (TP):

- ability to carry mass transport of liquid and gaseous goods,
- low cost of movement,
- very favorable offer in terms of transport time resulting from the huge throughput of pipelines and gas pipelines and the high reliability of supply,
- poor availability of space which is the result of a relatively rare and inconsistent network of pipelines and gas pipelines.

Combined transport (TKO):

- able to propose a combined transport offer with combined benefits of different modes of transport,
- opportunity to reduce the cost of movement while maintaining the required quality of transportation services,

- opportunity to reduce the damage and losses as well as handling and storage work through the use of containers and pallets,
- increased flexibility of supply by offering better spatial and temporal availability of services to more customers.

Bimodal transport (TB):

- allows the transport of container semi-trailers by rail, using their direct placement on suitably equipped wagons (elimination of the railway wagon),
- the car semitrailer is part of the train,
- can use different types of semi-trailers (closed box, open, tanks, refrigerators, etc.),
- possibility of "door to door" transportation with the use of rail and road,

Multimodal transport (TMD):

- "freight using at least two different modes of transport on the basis of the contract of multimodal carriage, from one country where the goods were acquired by a multimodal transport operator, to the marked location in another country".

Distribution (D) can be presented as follows:

$$D = D(I, KD) \tag{4.12}$$

where:
I—essential features of distribution:

- activities related to the movement of products from their place of manufacture to purchase (acquisition) by the final purchaser; the aim of distribution is to provide consumers and users with the products they desire in the appropriate time, place and terms with the lowest possible total cost of distribution,
- manufactured products are separated from their buyers by the barriers of place, time, range and property; overcoming them requires planning, organization and control over numerous transaction and property relations between manufacturers and final buyers; these take the form of streams of information, promotions, negotiations, product orders, receivables, property rights and risks, flowing through the distribution channels (KD); channels differ in terms of the type of participants carrying out activities related to these flows, as well as the length, width, and structuring of functions and the type of relationship integrating the activities of companies that make up the channel and support its functioning,
- decisions concerning the selection of the channel(s) of distribution (KD) are among the most important in marketing (…). The process of decision-making must take into account the correlation between distribution and other marketing instruments; after all, decisions on distribution are in fact determined by arrangements relating to the product, price and promotion on the one hand, while on the other hand, they may affect their formation due to the limited mobility of the distribution system;

KD—distribution channel:

- structured organizational elements of the company (sales and marketing departments, product storage, transport divisions, etc.) as well as external agents who sell the product on the market; a set of interdependent organizations involved in the delivery of products and services to their users or consumers, or
- a chain of links (institutions and individuals), through which one or more streams associated with marketing activities flow.

Based on the above, distribution channels (KD), can be described as follows:

$$KD = KD(KDK,\ KDP,\ KDU) \tag{4.13}$$

where:
KDK—distribution channel for consumer goods,
KDP—distribution channel for manufacturing goods,
KDU—distribution channel for services, and moreover:

$$KDK = KDK(B\ (Pr,\ NI),\ P(Pr,\ A,\ NI),\ P(Pr,\ D,\ NI),\ P(Pr,\ H,\ D,\ NI), P(Pr,\ A,\ H,\ D,\ NI)) \tag{4.14}$$

where:
B—direct distribution channel in which:

Pr—producer,
NI—individual buyer;

P—indirect distribution channel in which:

A—agents,
D—retailers,
H—wholesalers,

also

$$KDK = KDK(B(Pr,\ Nt),\ P(Pr,\ DS,\ Nt), P(Pr,\ A,\ Nt),\ P(Pr, A, DS, Nt)) \tag{4.15}$$

where:
DS—distributor;
Nt—institutional buyer;

and

$$\mathrm{KDU} = \mathrm{KDU}(\mathrm{B}(\mathrm{Pu}, \mathrm{Nu}),\ \mathrm{P}(\mathrm{Pu}, \mathrm{A}, \mathrm{Br}, \mathrm{Nu}),\ \mathrm{P}(\mathrm{Pu}, \mathrm{Fr}, \mathrm{Nu})) \tag{4.16}$$

where:
Pu—service provider;
Br— broker;
Fr—franchisee;
Nu—service buyer.

It is a system structure based on components. The "life" of a logistics storage system usually takes from 3^12 years (0—time as an independent variable of the process: hours, days, weeks, months, years), and is based on the processes (PR) implemented during that time. There are two basic types of processes: operational and control. The operational processes are energy-driven and produce tangible goods, while control processes are informative and interact with operational processes in a structured manner, in order to extract their maximum energy, power, efficiency, etc.

The concept of processes is defined as a sequence or partially orderly way of action that share a common goal, integrated in terms of time, energy consumption, costs and the assessment of the quality of workmanship. The components, which explicitly or indirectly make up logistic processes include:

- people,
- material goods,
- capital,
  Information and
- environment.

Logistics, as the coordination of processes, treats processes in an integrated way, which should be coordinated and run smoothly. The implementation of the logistics process requires coordination between its components in both time and space. Proper planning of the process therefore requires a timetable, indicating the place of origin and receipt for each of the components of the process, and the people responsible for the proper implementation of these activities.

Reflections on the concept of the logistics process allow for presentation of the relations from the previous pages of the monograph in the form of the following expression:

$$\mathrm{S}(\mathrm{PR}(\theta)) = \mathrm{SL}(\mathrm{Z}(\mathrm{PRz}(\theta)),\ \mathrm{P}(\mathrm{PRp}(\theta)),\ \mathrm{M}(\mathrm{PRm}(\theta)), \mathrm{T}(\mathrm{PRt}(\theta)), \mathrm{D}(\mathrm{PRd}(\theta)), \mathrm{R}) \tag{4.17}$$

The left side of the expression above shows the implementation of processes in the logistics system, and the right side—the implementation of processes in specific components of the logistics system, wherein:

$$\mathrm{PR}(\theta) = \mathrm{PR}(\mathrm{PRZ}(\theta)), \mathrm{PRP}\ (\theta), \mathrm{PRM}\ (\theta), \mathrm{PRT}\ (\theta), \mathrm{PRD}(\theta) \tag{4.18}$$

or

$$\mathrm{PR} = \mathrm{PR_Z} \cup \mathrm{PR_P} \cup \mathrm{PR}_M \cup \mathrm{PR}_T \cup \mathrm{PR}_D \tag{4.19}$$

The implementation of processes (PRz(θ)) in the component of a logistics system (SL) which involves supply (Z) can be expressed as follows:

$$\mathrm{Z}(\mathrm{PRz}(\theta)) = \mathrm{Z}(\mathrm{PRz}(P^1 z(\theta), P^2 z(\theta), \ldots, P^1 z(\theta)) \tag{4.20}$$

where:
$P^1z(\theta)$—demand for raw materials, materials, components, assemblies, … in time θ
$P^2z(0)$—scheduling distribution … in time θ
$P^3z(0)$—planning supply schedule … in time θ
$P^4z(0)$—maintaining inventory for the smooth production in time θ

For processes PRm(0) in the component, warehouses (M) are presented as follows:

$$\mathrm{M}(\mathrm{PR_M}(\theta)) = \mathrm{M}(\mathrm{PR_M}(p_M^1(\theta), p_M^2(\theta), \ldots, p_M^1(\theta)) \tag{4.21}$$

where:
p_M^1 (θ)—acceptance (delivery, unloading, sorting, unpacking, maintenance)…in time t, θ
p_M^2 (θ)—storage (placement, maintenance) …in time, θ
p_M^3 (θ)—picking (retrieval, reloading, packaging, transport to the release area)… in time, θ
p_M^4 (θ)—release (shipment, loading)… in time t, θ

Other processes are defined in the same way as the previous two. Based on the described formal model, the principle is shown for the assessment of energy intensity of LSM using a calculation model: boundary conditions for the organization of LSM (Fig. 4.9): the warehouse, the work, the equipment, were assumed at:

- Equipment 1,2, …, m
- Docks (in-out gates for freight units) 1, 2, 3, …, s
- Storage areas (shelving slot/storage area) 1, 2, 3, …, k

The storage process consists of: the process of admission to the warehouse, storage, picking, and release. The model assumes that picking is performed in pallet units. Thus, we can distinguish the following stages, which a freight unit goes through from its placement near the dock gate, through the process of storage to release from stock:

Pallet retrieval (i),
Pallet transport (j),
Deposition in selected storage area (l).

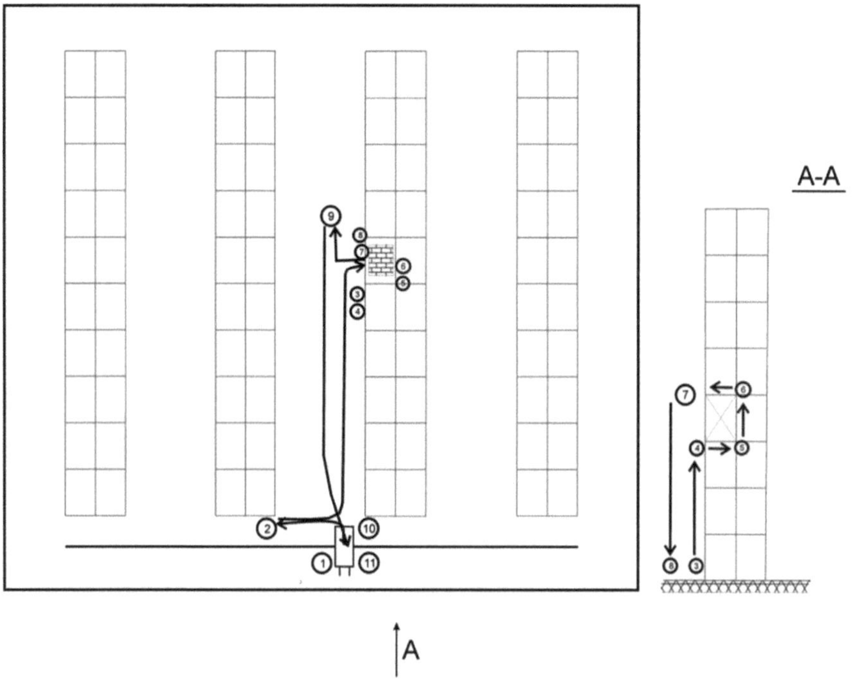

Fig. 4.9 Line diagram of a forklift truck for an Case terminal with a goods lock: *1*—SKU retrieval and turn backwards; *2*—transport to the trailer; *3*—placing the SKU; *4*—relapse; *5*—the empty truck goes back to the point of handover

In special cases, a change of the transport device occurs between stages i, j, l. Each stage is carried out by a device, i.e. storage of pallets (movement of the freight unit in the transport and storage system) can be expressed as three integers (i, j, l) specifying the number of the device used for storage.

Storage consists of the following operations:

O_1—Driving an empty forklift from standby area to the freight unit,
O_2—Lifting,
O_3—Delivery of pallets to PP_1,
O_4—Empty forklift returns to M (standby area),
O_5—Reloading,
O_6—Transport from PP_1 to PP_2,
O_7—Reloading,
O_8—Delivery of pallet to rack,
O_9—Lifting,
O_{10}—Empty forklift returns to standby area of the device M.

For each device, the following data are given:

Specific energy intensity in horizontal movement,

Specific energy intensity in vertical movement.

These data are stored in a ZE matrix sized $m \times 2$, defined as follows:

$$ZE_{i1} = \begin{cases} > 0 & \text{unit ZE i---of that device in horizontal movement} \\ 0 & \text{if the device does not carry out movement} \end{cases}$$

$$ZE_{i2} = \begin{cases} > 0 & \text{unit ZEi---of that device in vertical movement} \\ 0 & \text{if the device does not carry out movement} \end{cases}$$

$$ZE = \begin{matrix} 1 \\ 2 \\ . \\ . \\ m \end{matrix} \begin{bmatrix} ze_{11} & ze_{12} \\ ze_{21} & ze_{22} \\ . & . \\ . & . \\ ze_{m1} & ze_{m2} \end{bmatrix} \quad (4.22)$$

4.2 Unloading, Admitting Freight Unit to the Warehouse

For the set of n pallets located in the selected handling dock and with known: mass, its storage location in the warehouse.

A matrix of PT transport parameters is built for the set, sizing $n \times 5$, where:

PT_{i1} = mass of the i-th pallet,
PT_{i2} = unloading stage path for the i-th pallet,
PT_{i3} = transport stage path for the i-th pallet,
PT_{i4} = placement stage path for the i-th pallet,
PT_{i5} = storage height of the i-th pallet,

$$PT = \begin{matrix} 1 \\ 2 \\ 3 \\ . \\ n \end{matrix} \begin{bmatrix} pt_{11} & pt_{12} & pt_{13} & pt_{14} & pt_{15} \\ pt_{21} & \cdots & \cdots & . & pt_{25} \\ . & . & . & . & . \\ . & . & . & . & . \\ pt_{n1} & . & . & . & pt_{n5} \end{bmatrix} \quad (4.23)$$

For each pallet (freight unit) should have a given (defined) technology of unloading. As transshipment from one handling device to another may occur between stages, one must specify the specific energy consumption of these operations (the matrix size $m \times m$).

As three numbers, each of which is the number of executing the particular stage of unloading (i, j, k), for Case (3, 5, 1). These three number mean that for a specific pallet, the first stage is implemented by device #3, the second stage is carried out by

the device #5 and the third—by the device #1. The set of all these technologies for the entire set of pallets makes up the matrix; it is denoted as T.
where:

tij—means the number of the device carrying out the j − t stage of unloading the pallet i.

i = 1, 2, 3, …, m

j = 1, 2, 3

$$T = \begin{bmatrix} t_{11} & t_{12} & t_{13} \\ t_{21} & t_{22} & t_{23} \\ t_{n1} & t_{n2} & t_{n3} \end{bmatrix} \tag{4.24}$$

We shall mark Z, which will be defined as follows:

$$Z_{ij} = \begin{cases} > 0 & \text{when the reloading from the one device i to the second device j is possible} \\ & \text{and is equal to the single energy consumption of this operation} \\ 0 & \text{there is no reloading} \end{cases}$$

$$Z = \begin{bmatrix} Z_{11} & Z_{12} & \cdots & Z_{1M} \\ Z_{21} & Z_{22} & \cdots & Z_{2M} \\ \cdots & \cdots & \cdots & \cdots \\ Z_{M1} & \cdots & \cdots & Z_{MM} \end{bmatrix} \tag{4.25}$$

For each pallet and with the taking into the consideration (T) technology, it is easy to determine elementary energy consumption. This energy consumption is going to be indicated as a matrix O with the dimension n × 10. Elements of this matrix are defined below:

$$O_{ij} = \begin{cases} 1 & \text{single energy consumption which is need to carry out an operation of j detail} \\ 0 & \text{if the operations j is not realized} \end{cases}$$

This matrix is determined on the basis of matrix ZE, Z and T.

$$\begin{aligned}
O_{i1} &= ZE(T(i,1),1) \\
O_{i2} &= ZE(T(i,1),2) \\
O_{i3} &= 0_{i1} \\
O_{i4} &= 0_{i1} \\
O_{i5} &= ZE(T(i,1),T(i,2)) \\
O_{i6} &= ZE(T(i,2),1) \\
O_{i7} &= Z(T(i,2),T(i,3)) \\
O_{i8} &= ZE(T(i,3),\ 1) \\
O_{i9} &= ZE(T(i,3),2) \\
O_{i10} &= 0_{i8}
\end{aligned}$$

To obtain energy consumption for each operation, it is necessary to correct elements of matrix, with taking the track, the load and the height into consideration. Coefficients of corrections are determined on the basis of the PT matrix. We are obtaining a matrix WK (n × 10), which is defined as a matrix of the coefficients of correction of energy consumption for each operation and for each pallet.

After the operation of matrix multiplication the matrix WK and matrix O, we are going to obtain the matrix E, where the elements of this matrix correspond to the energy consumption for each operation and for each pallet.

Matrix multiplication is defined as a:

$$WK_{n\times 10} \otimes 0_{n\times 10} = E_{n\times 10} \tag{4.26}$$

where:

$$e_{ij} = WK_{ij} \cdot O_{ij} \tag{4.27}$$

The total consumption of energy amount to:

$$E_C = \sum_{i=1}^{n} \sum_{j=1}^{10} e_{ij} \tag{4.28}$$

Important indicator values are determined based on the relations shown in the following sections of this paper; this applies to calculating the energy intensity of equipment for vertical and horizontal movement, and information subsystems. Moreover, one should take into account the share of energy intensity of the facility's lighting, the operation of office equipment and heating/cooling. The procedure is shown in block form in Fig. 4.10.

Heating/cooling energy is taken into consideration in the case of LSM, where there is temperature regime. In facilities such as cold storage, freezers, air-conditioned warehouses, the featured model takes into account this energy, but due to the lack of experimental data, it has not been fully tested for these cases.

In such cases, the model included the assessment of energy intensity of LSM: the heat penetrating through walls, ceiling and floor of the storage compartment, the heat dissipated from the refrigerated goods, the heat delivered by air, which was introduced into the chamber unintentionally, the heat associated with the work of the refrigeration unit, heat generated by people; and others. This situation is schematically shown in Fig. 4.11, and formally represented in the formula below:

$$Q = Q_d + Q_w + Q_L + Q_V + Q_{ah} + Q_{Ma} + Q_{Me} + Q_S \tag{4.29}$$

where:

Q_d—the heat penetrating through the walls, ceiling and floor of the cold room;
Q_w—the heat removed from the merchandise;
Q_L—the heat given up by the air unintentionally brought into the cold room;

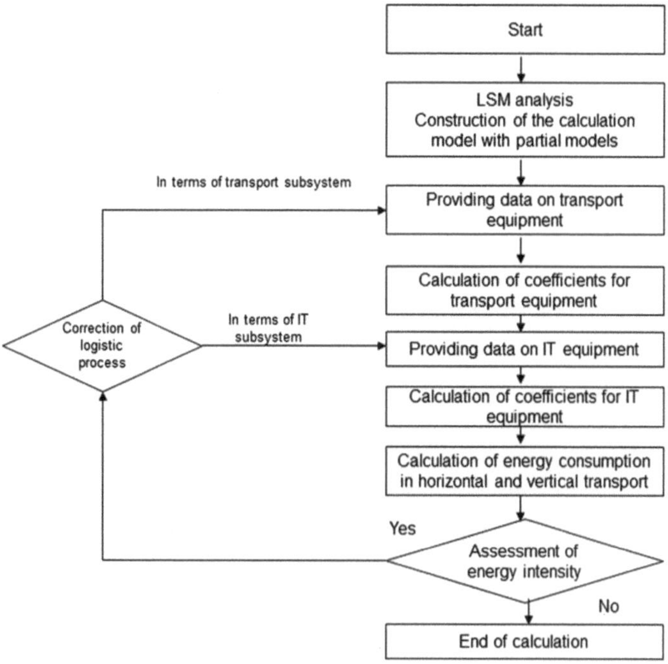

Fig. 4.10 Algorithm for the evaluation of energy intensity. Author's own work

Fig. 4.11 Energy balance in a sample store [273, 274]

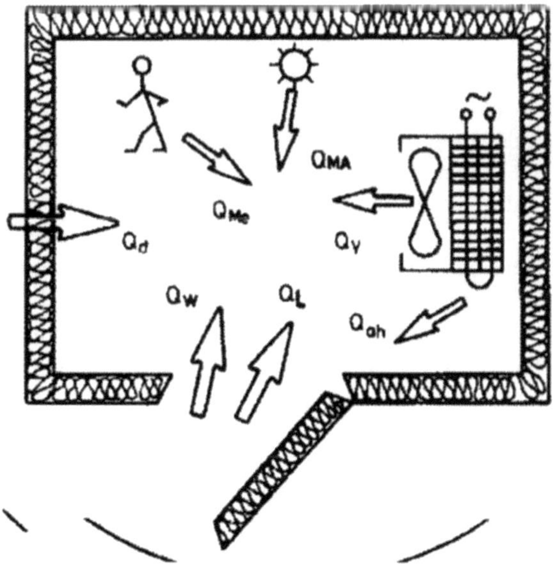

Q_V—the heat generated by the operating air cooler fan;
Q_{ah}—if need be, the heat generated during defrosting;
Q_{Ma}—the heat emitted by the lighting, the machines and similar equipment in the store;
Q_{Me}—the heat emitted by people;
Q_S—the heat constituting a standby in case of unforeseen changes in the store's thermal load.

4.3 Vertical Displacement Energy

It is assumed that the floor of the warehouse is the reference level, and any movement of the freight unit by a device in a vertical plane changes its energy intensity of passage through LSM, thus increasing it by the energy used or reducing by the energy recovered (e.g. in the case of placing a freight unit on a shelf, and then moving it again) the physical source is the potential energy (4.30).

$$Ep = W \cdot I \tag{4.30}$$

The potential energy is equal to the work to be done in order to move the palette and material vertically. Vertical movement of cargo in LSM is done by the lifting system, which is used to grip and raise/lower the freight unit vertically. In forklift trucks (like in pavers), this mechanism consists of a telescopic frame, a carriage, a carriage lifting mechanism and a frame tilting mechanism. The telescopic frame consists of two main parts: the outer fixed frame 8 and the inner mobile frame 9 (Fig. 4.12) moving up and down relative to the outer frame. The outer stationary frame of the lifting mechanism is pivotally attached to the body frame supports, between the drive shaft wheels with two axes 23. The upper ends of the vertical guides have two brackets mounted on the inside, with the pulleys 11 to guide the inner sliding frame. Rail guides 7 are welded on the inside of the clamps in vertical guides, providing the rails for rollers in the mobile frame.

The mobile frame 9 consists of two rails made of U-sections, joined at the top with a bar. At the bottom of each guide, two pulleys 6 and one disc 4 are mounted on roller bearings, bolted to the U-section using bolts. The pulleys 6 support the mobile frame along the longitudinal axis of the truck and protect against slanting, and the rollers 11 and 4 placed on the fixed frame and mobile frame support the mobile frame parallel to the axis of the drive wheels; they also protect it against bending.

The carriage 1 for attaching the working equipment and for gripping cargo consists of two side stringers connected with crossbars. Pulleys 2 and 5 of the carriage are mounted on the stringers, they roll along the rails welded on the inside of the clamps of the mobile frame vertical guides. The fork is fixed to the carriage's crossbars using brackets 17 and 18, and the protective frame 15 is attached using bolts.

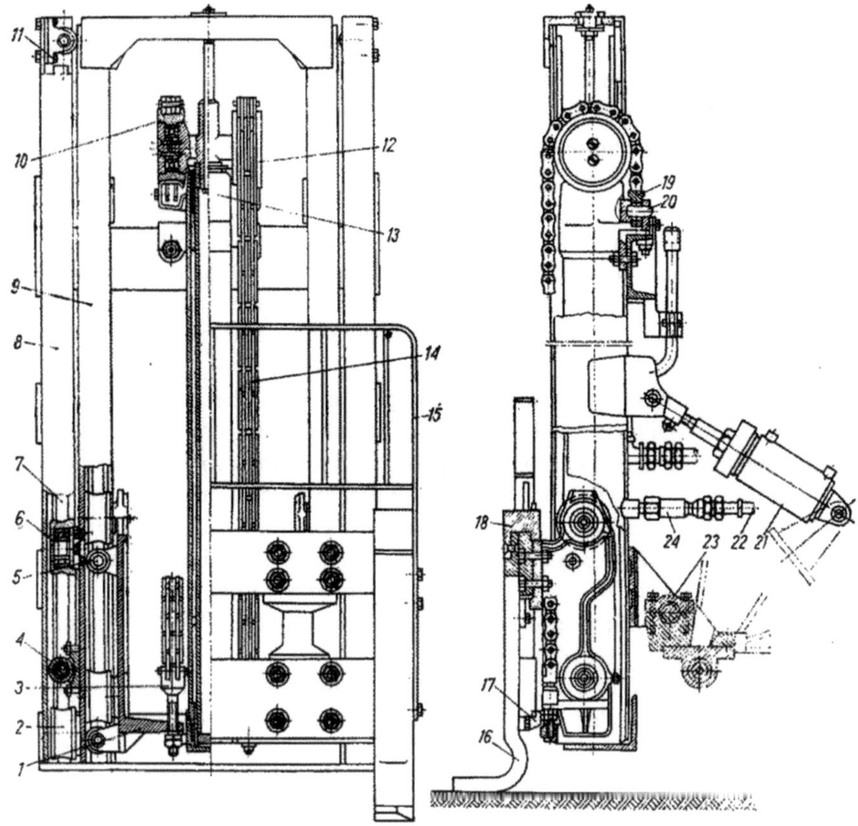

Fig. 4.12 Diagram of the forklift's lifting mechanism

The truck's lifting mechanism consists of plate chains 14, a bracket 12 with guiding pulleys, and a hydraulic lifting cylinder 13, single-acting plunger type. The plate chains are fed through pulleys mounted on the wheel bracket on roller bearings. One end of the 3chains is attached to the carriage, and the other—to the two-armed compensating lever 19, mounted on the axis 20 in the lifting cylinder head, which provides uniform load on both chains in the case of skewing the carriage.

The carriage is lowered by the weight of the load and the moving parts of the lifting mechanism; the plunger slides in said cylinder, pumps the oil into the tank through the throttle. The lower end of the cylinder is secured by a shackle to the lower crossbar of the fixed frame. The working fluid (e.g. spindle oil) is passed from the distributor to the lifting cylinder through the hose 22 and the throttle 24.

The tilting mechanism of the frame consists of two tilt cylinders 21, piston type and double-acting. The eyes of the tilting cylinders are pivotally attached to the body frame brackets and the rods with tips screwed onto them—to the brackets of the fixed frame of the lifting mechanism. The working fluid from the hydraulic distributor is fed by connectors to the top or bottom of the cylinder, which moves in the piston rods and pistons, tilting the telescopic frame forward or backward.

The hydraulic system (Fig. 4.13) consists of a vane double-action hydraulic pump, a hydraulic distributor with pressure reducing valve, hydraulic actuators (cylinders), working fluid container and conductors. This system operates as follows. The oil is poured into the tank through the filtered inlet 3. The pump 1 draws the oil from the tank through the hose 2, and pumps it through line 13 to the distributor valve 12.

The oil flows from the distributor through line 11 to the single-acting lifting cylinder 14, through lines 6 and 7—to the tilt cylinder 5, and through the lines 8 and 9—to cylinders (actuators) 10 of the work tools. From the cylinders (actuators) of the work tools, the oil is returned to the tank through the distributor and hose 4. The truck's control mechanism consists of an adjuster control pedal, a hydraulic brake control pedal, a handbrake, a direction of travel lever, the steering column, ignition and a control lever for cylinders (actuators) of the work tools, tilt and lifting.

The frame of the lifting mechanism may be tilted forward 3°–5° and backward 10°–15° (Fig. 4.14).

The hydraulic drive power of forklift trucks with hydraulic actuators powered by a single pump is calculated based on the maximum force of the lifting mechanism

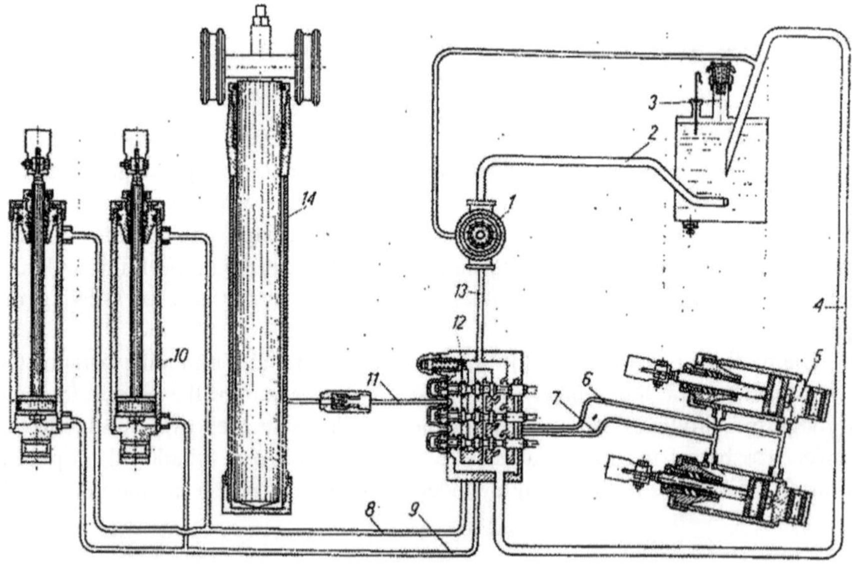

Fig. 4.13 Diagram of the forklift's hydraulic system

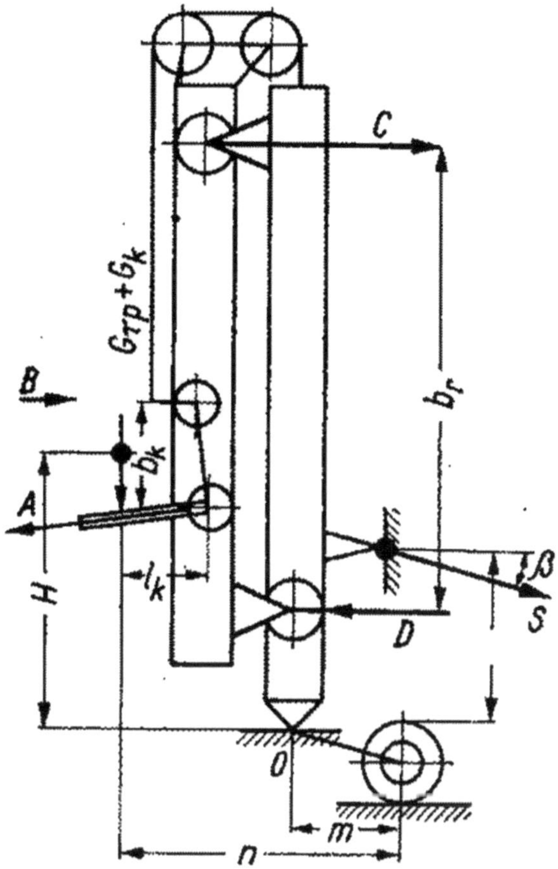

Fig. 4.14 Diagram for the calculation of forklift lifting work

actuator, because the connection with the work of lifting hydraulic cylinders for other operations is impossible.

The necessary pump motor power:

$$N = \frac{\sum W \cdot \upsilon}{2 \cdot 1000\eta_h \cdot \eta_p} \tag{4.31}$$

where:

$\sum W$—the work of lifting the carriage with the fork, taking into account the friction in the guides of the carriage and the moving frame [N],

υ—load raising speed (m/s),

η_h—actuator efficiency (0.8 is assumed),

η_p—pump efficiency (0.8 is assumed).

The lifting work is defined by formula (4.32).

$$\sum W = \frac{2(G_l + G_k)}{\eta_b} + G_r + W_t \quad (4.32)$$

where:
G_l—the force of gravity (weight) of the load [N],
G_k—the force of gravity of the carriage and the working fittings [N],
G_r—the force of gravity of the inner frame [N],
W_t—the work of the friction forces during the movement of the carriage and the moving frame of the lifting mechanism [N].

In order to calculate the frictional resistance one should determine the support reaction to carriage rollers A and B (Fig. 4.11) and moving frame rollers C and D:

$$A = B = (G_l + G_k)\frac{l_k}{b_k} \quad (4.33)$$

$$C = D = (G_l + G_k + G_r)\frac{l_r}{b_r} \quad (4.34)$$

where:
l_r, l_k—the distance from the centre of gravity of respectively the load with the carriage and the carriage with the moving frame to the load gripping point (l_r can be assumed to be equal to l_k),
b_r, b_k—the spacing of the bumper rollers of respectively the carriage and the moving frame.

Then:

$$W_t = \left[2(G_l + G_k)\frac{l_k}{b_k}\left(\frac{fd}{D} + \frac{2\mu}{D}\right) + 2(G_l + G_k + G_r)\frac{l_p}{b_p}\left(\frac{fd_1}{D_1} + \frac{2\mu}{D_1}\right)\right]k\ [\mathrm{N}] \quad (4.35)$$

where:
d and d_1—the diameters of the hubs of the frame carriage bumper rollers,
D and D_1—the diameters of the bumper rollers of the carriage and the frame,
F—the coefficient of friction in the hubs,
μ—the coefficient of friction of the rolling of the rollers on the guides,
K—a coefficient representing other additional resistances (the type of wheels and their bearing system, 1.1–1.3 is assumed).

Force S on the frame tilt actuator piston rod is calculated from the equation of the moments of the forces acting on the frame relative to point 0 (i.e. the point of hinged fixation), as it tilts forwards at angle α at the upper position of the moving frame and the carriage with the load.

$$S = \frac{(G_l + G_k + 0.75G_{rg})H_{\max} tg\alpha + G_l(n - m)}{h\cos\beta}[N] \tag{4.36}$$

where:
G_{rg}—the force of gravity (weight) of the main frame, including the tilt actuator,
0.75 and 0.25—coefficients approximately indicating the location of the centres of gravity of the frames,
H_{max}—the maximum carriage-lifting height.

The dependence between the actuator piston rod force and the hydraulic actuator parameters can be expressed by formula (4.37) when the piston rod is inside the cylinder and by formula (4.38) when the piston rod is not inside the cylinder.

$$S = \frac{\pi(D_c^2 - d_t^2)p\eta_m}{4} \tag{4.37}$$

$$S = \frac{\pi D_c^2}{4}p\eta_m \tag{4.38}$$

where:
D_c—the cylinder diameter (cm),
d_t—the piston rod diameter (cm),
p—pressure (N/cm^2),
η_m—the actuator efficiency (0.95 is assumed).

The working liquid consumption when the piston is inside the cylinder, taking into account losses, amounts to:

$$V = \frac{\pi(D_c^2 - d_t^2)v_t}{4\eta_c} \quad (cm^3/min) \tag{4.39}$$

$$V = \frac{\pi D_c^2 v_t}{4\eta_c} \quad (cm^3/min) \tag{4.40}$$

where:
v_t—the piston speed (cm/min),
η_c—the volumetric actuator efficiency (approximately equal to 1 when rubber and leather seals are used).

Forklift trucks incorporate hydraulic pumps with a delivery of about 80 l/min at about 1700 rpm and a working pressure of 1000 kN/m^2 for mast raising/lowering and forwards/backwards tilt.

In order to translocate the unit load in the vertical in LWS, energy must be supplied to the latter. In the case of a forklift truck or a storage and retrieval machine, taking out the unit load from a given layer larger than a unit requires energy for the raising of the empty fork carriage while the lowering of the carriage necessitates the introduction of a force braking the carriage with the unit load, ensuring the safe lowering of the latter to the required level. Depending on the size of the forklift truck, the carriage weighs 100–300 kg.

In such cases, technical systems making it possible to recover the remaining part of the potential energy of the forklift carriage and the unit load are used. Thanks to energy recovery LWS energy consumption and the energy consumption index of unit load passage through LSW are reduced.

Therefore pallet lifts and S/R machines (in some warehouse zones) are introduced into LSWs, for which the power demand can be defined by respectively formula (4.41) and (4.42).

$$N_s = \frac{kG_l v}{1000\eta} \quad \text{(kW)} \tag{4.41}$$

$$N_u = \frac{\sum W \cdot v}{1000\eta} \quad \text{(kW)} \tag{4.42}$$

where:
G_l—the force of gravity of the load (N),
V—the carriage (platform) lifting speed (m/s),
η—the efficiency of the device mechanisms,
K—a coefficient representing the influence of the counterweight, amounting to 04–06, when there is no counterweight, k = 1,
$\sum W$—the S/R machine driving resistance on the track (N), defined as for a runway truck.

These infrastructural factors also affect the changes in the energy intensity indicator, which in this case depending not only on the payload but on the forklift as well.

The energy recovery systems, which are mentioned above are very slowly yet systematically introduced into the standard equipment of handling machinery and equipment. The literature describes the case of taking into account the recovery of potential energy [151] (or in other papers such as [80]) by regenerative braking using an electric motor, and the accumulation of energy in the battery, while the classic solution requires feeding it to the system.

4.4 Horizontal Displacement Energy

In the LSM, the movement of the freight unit horizontally is carried out using forklifts and conveyors. Literature describes the issue of the theory of traffic: papers speaks of motor vehicles [11, 252], while [302] touches upon rail vehicles. A forklift belongs in the handling vehicles group; it travels at a speed of 5–7 km/h, has an unladen weight depending on the capacity parameters, upon smooth surface. Is characterized by a varying total mass in subsequent cycles. These vehicles can be found as front carts, side cars etc. They feature a various number of wheels and can serve as tow trucks.

In the literature one can find such papers as [155], in which the authors derive the equation for traffic of tractors, loaders and agricultural machines, i.e. the movement of vehicles which more closely resembles the movement of a forklift.

Based on their own experience and literature studies, the author derived an equation for forklift movement. This equation allows calculating the kinetic energy needed to overcome the resistance to movement of FN (V) for each forklift.

Starting with the simplest case, it can be expressed as follows:

$$F = m \cdot a + m \cdot g \cdot \mu \tag{4.43}$$

wherein $\mu = f(mat1, mat2)$.

The equation of motion of the vehicle can be expressed by Eq. (4.36).

$$F(V) = m \cdot a + W_t \tag{4.44}$$

$m \cdot a$ Forces of inertia
$w = f_t \cdot Q_t$ Friction
f_t Rolling resistance coefficient
$F(V)$ Drive forces

Resistance to motion of the vehicle can be divided into: primary and secondary. The first group includes rolling resistance, damping, toe resistance and aerodynamic resistance; and the secondary resistances include: climbing resistance, steering resistance, starting and traction resistance.

Rolling resistance depends on: chassis structure, surface type, type of tires, tire pressure (in the case of pneumatic wheels) and the operating status of the vehicle (laden/unladen).

$$\begin{aligned} &\sum x = 0; \quad F_k - X = 0 \\ &\sum y = 0; \quad Z - Q = 0 \\ &\sum M^o = 0; \quad Z \cdot e - F_k \cdot r_k = 0 \end{aligned} \tag{4.45}$$

$$F_k = Z\left(\frac{e}{r_t}\right) \tag{4.46}$$

$$F_k = Q \cdot \frac{e}{r_t} \tag{4.47}$$

$$W_t = Q_p \cdot f_{tp} + Q_T \cdot f_{tt} \tag{4.48}$$

Damping resistance is approximately $r_{tz} = 5\ \%$ of the rolling resistance.
Toe resistance, $\beta = 0 \div r_{zk} = 5\ \%$ of the rolling resistance.

Additional resistances include:

Riding uphill,
Driving round bends,
Towing.

Resistance for riding uphill

W_i = *drag lift.*

$$F_{st} = Q \cdot \sin\alpha \tag{4.49}$$

$$F_{st} = m \cdot g \cdot \sin\alpha \tag{4.50}$$

where $\alpha = \sin\alpha = tg\alpha$ up to 10°.

Resistance in bends—excel, calculations Towing resistance:

$$W_{uciagu} = W_{tocz.poj.d} + W_{i.pj.docz} + WSK_{poj.docz.} \tag{4.51}$$

Movement equation:

$$\xi \cdot m \cdot \frac{dV}{dt} + W_{r.coal} = F_N(V) \tag{4.52}$$

where:
ξ—rotating masses coefficient,

On the basis of experience, the paper assumes that horizontal transportation—gravitational and powered roller conveyors—make up the vast majority in LSM. The model can be extended to others (see Fig. 4.15), such as suspended conveyors, belt conveyors, etc.

A roller conveyor (Fig. 4.16) consists of a straight Sect. 3 with a length of 2.0^2.5 m curved 1, swivellable 2 and rotating 4. Straight shafts 6 are mounted on ball bearings. Transferring the load from one conveyor to another involves steering rollers 5. These use pretty much the same diameter as the rollers in belt conveyors. The length of the rollers is determined by the width of the cargo layer, and the spacing between them should be slightly smaller than half the length of the conveyed items. Recently used are light straight rolls of plastic or rolls consisting of

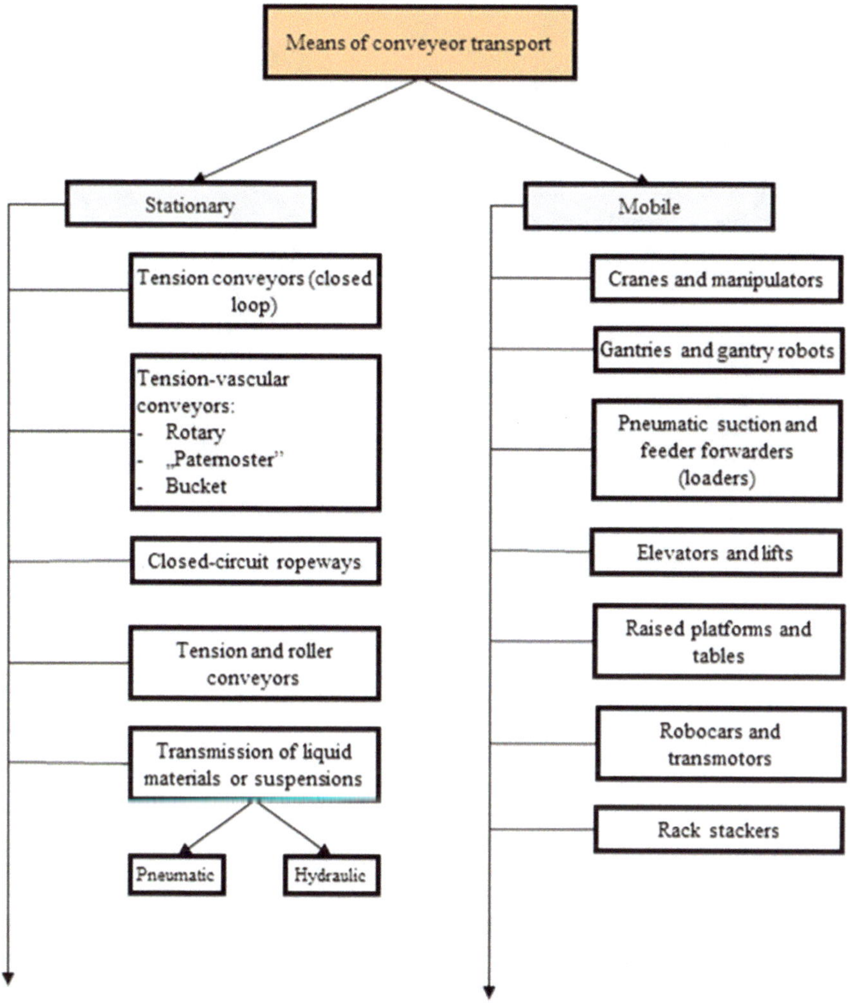

Fig. 4.15 Breakdown between conveyors [135]

individual charts. The ability to move the load at a small angle is achieved by the sliding resistance of load being mostly converted to rolling resistance.

Non-powered roller conveyors are manufactured in the form of sections with rollers arranged in lightweight horizontal or slightly inclined frames (2–3°).

When the load weight is G, spacing between the rollers l_r, the distance between the items carried l_l, then the weight per meter is given by the formula:

$$q = \frac{G}{l_t} \quad \text{(kg/m)} \tag{4.53}$$

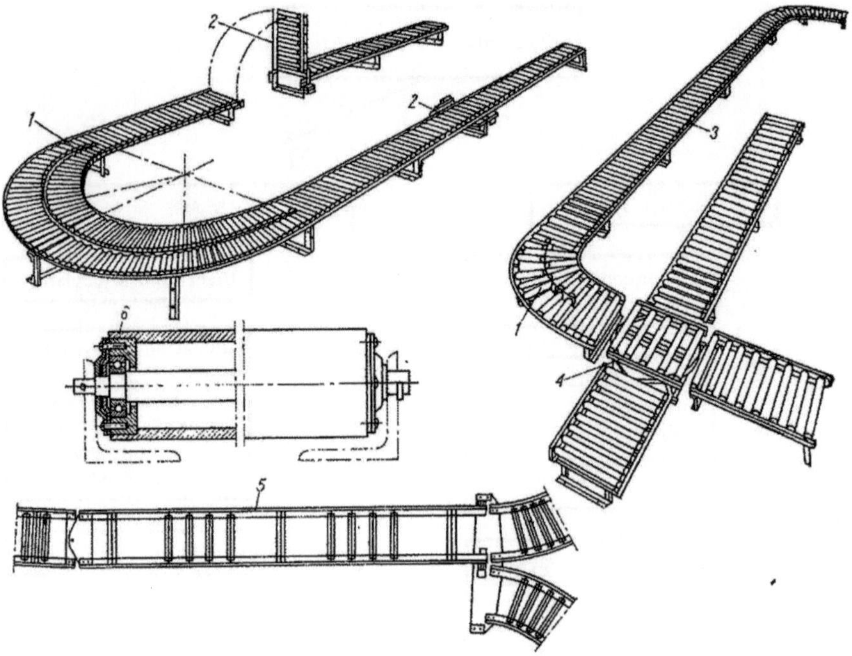

Fig. 4.16 Roller conveyor. Author's own work

and the weight per one roller is:

$$q_r = ql_r = \frac{G \cdot l_r}{l_t} \, (\text{kg}) \tag{4.54}$$

If the weight of the shaft q_0 in kilograms, the diameter of the shaft D_r, the diameter of the spigot d and the coefficient of friction f, then the resistance of movement of a load on surface inclined at an angle p, resulting from frictional forces on the pin and the component of the load's and the shaft' weight brought to the roller's surface:

$$W = q_r g f \frac{d}{D_r} \cos \beta \pm q_r g \sin \beta + q_o g f \frac{d}{D_r} \quad (\text{N}) \tag{4.55}$$

In the case of cargo movement down, the second part has a minus sign, and in the case of upward movement—a plus sign. The rotation resistance of the shaft due to the friction between the drive shaft and the load:

$$W_t = q_r g \mu \cos \beta \quad (\text{N}) \tag{4.56}$$

The total rotation resistance of one shaft:

$$W = W_c + W_t = q_r g\left[\left(f\frac{d}{D_r} + \mu\right)\cos\beta \pm \sin\beta\right] + q_o g f\frac{d}{D_r} \quad \text{(N)} \tag{4.57}$$

The sum of the rotation resistances of all the rollers:

$$\sum W = qgL\left[\left(f\frac{d}{D_r} + \mu\right)\cos\beta \pm \sin\beta\right] + q_o g f\frac{d}{D_r}\cdot\frac{L}{I_r} \quad \text{(N)} \tag{4.58}$$

where:
L—conveyor length
μ—coefficient of rolling friction

The engine power for a driven roller conveyor is defined by the formula:

$$N = \frac{\sum W \cdot v}{1000 \cdot \eta} \quad \text{(kW)} \tag{4.59}$$

where:
v—the speed of the load's movement
η—gear efficiency.

Resistance due to friction against the movement of one object for a non-driven conveyor can be determined using the formula:

$$W_l = \frac{\sum W \cdot l_m}{L} \tag{4.60}$$

where:
l—the length of the freight unit transfered.

In order to reach the overall resistance, one should add the resistance arising from the inertia of the rollers and of the cargo, since in this type of conveyor, rollers often stop during the intervals between the conveyed objects. If the shaft's moment of inertia Ir and its angular acceleration—then the resistance caused by inertia can be determined using the formula:

$$W_b = \frac{4I_r \cdot l_m}{D_r^2 l_r} j + Gj \tag{4.61}$$

Assuming that not all the weight is concentrated on the surface of the roller, the moment of inertia of a shaft with a mass of q_o, can be assumed as:

$$I_r = \frac{q_o D_r^2}{5.25} \tag{4.62}$$

and linear acceleration:

$$j \approx (0.5 \div 1.0) \quad (\mathrm{m/s^2}) \tag{4.63}$$

The mass of the shaft may be approximately determined based on analyzing the given structure data, as:

$$\mathrm{q} = 10\mathrm{KBr} \quad (\mathrm{kg}) \tag{4.64}$$

where:
Br—shaft length (m)
K = 1 or K = 2—coefficients taking into account the type of rollers (normal, heavy-duty).

The total movement resistance of cargo will be:

$$\mathrm{W} = \mathrm{W_l} + \mathrm{W}_b \tag{4.65}$$

Knowing the resistance W, one can specify the drag coefficient w and the equal tangent of the slope of the conveyor:

$$w = tg\beta = \frac{\mathrm{W}}{\mathrm{G}} \tag{4.66}$$

4.5 Energy Intensity of IT Subsystem

The energy intensity of the information interchange subsystems and their impact on the energy consumption of LSM is rarely taken into account in the literature. Below is the way to calculate the energy intensity of information systems and their impact on energy consumption in LSM.

An analysis of case studies available in the literature allows estimating the energy intensity of IT subsystems at not more than 8 % of the energy intensity of LSM. This is assumed to be the maximum value. If this value is exceeded, such a case should be treated individually. The rest of the paper discusses the subsystems separately; EDI information and automatic identification, and in case of exceeding the limit of energy intensity, subsystems must be checked individually.

LSM information subsystems typically consist of a subsystem for automatic identification, documents interchange (traditional or electronic EDI), or WMS computer systems.

The heart of the IT subsystem is the database system, most commonly integrated with the WMS (*Warehousing Management System*).

The result of the IT subsystem in LSM includes the required documents such as invoices, GM, VP, etc.

Automatic identification is involved in barcode marking goods, storage areas, documents, employees (IDs with barcodes if necessary, or biometric identification). Labeling requires printed barcode symbols to be placed in appropriate places. The encoded information is read using barcodes, verified with the help of scanners.

Barcodes can be replaced with radio codes, commonly designated RFID, which in some specific circumstances replace barcodes. Despite the high cost of both the purchase and operation of this technology, they are used in some LSMs.

Warehouse services widely use portable terminals. These devices are made of a portable computer and a scanner with the function of storing and transmitting data (complex versions also allow processing data). Data collection is done by reading barcode with the barcode scanner, entering them using the keyboard, or sending the message from other sources of information, e.g. a computer. The terminals can be used in warehouses as separate devices or as operator panels. The current terminals have widely developed software to analyze the data collected, depending on their memory size (which is usually limited). Sorting, error checking, etc. Due to their mobility, portable terminals have low weight and are battery-poweed. They feature co-operation with other devices (printers, readers) through appropriate wired or wireless interfaces [152].

Radio terminals have the ability to process and transfer data to and from the on-line database server. In order for the terminal to function properly, the work area must be within range of the radio antenna (access point). This coverage refers to the entire working space horizontally and vertically. Portable radio terminals have similar functionalities as PDAs or laptops. This includes access to the Internet—i.e.

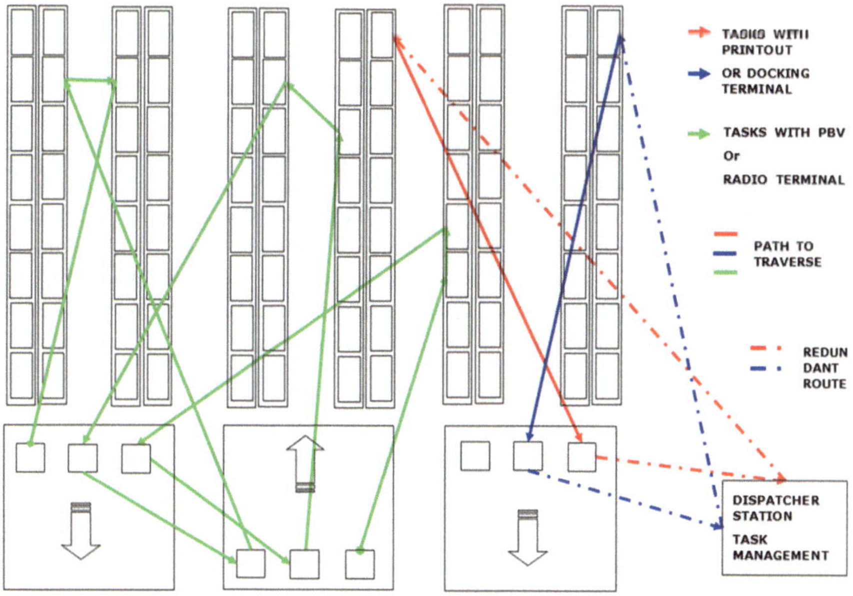

Fig. 4.17 Summary of the methods of exchanging information with a forklift operator. Author's own work [310]

a web browser and e-mail. This is used for customer service and employee management, e.g. assigning tasks between employees, monitoring their work, queuing tasks for employees, etc.

The Figure illustrates the relationship between the distance traveled by a forklift and the method of information interchange (Fig. 4.17).

4.5.1 Automatic Identification

Reference [157, 297] characterized the automatic identification technologies used in LSM: one-dimensional (1D), two-dimensional (2D) or radio (RFID—*Radio frequency identification*).

Breakdown of barcode readers by mobility:

- Portable,
- Stationary.

Breakdown of barcode readers by their way of readout:

- Laser—emitting a laser beam,
- Diode—emitting a light beam,
- Vision—scanning the image as in video cameras

The above-mentioned radio technologies can be used to read the current position of a freight unit, a forklift truck or an employee in the LSM. Thus, it becomes possible to assign tasks to staff and forklift trucks from the closest location. Upon the execution of the order, equipment can be directly routed to the next task, or directed to "standby". This is an implementation of the recommendations, which can be found in papers: [78, 79, 147] and others, regarding the minimization of empty runs of handling vehicles involving the elimination of simple work cycles in favor of complex cycles.

Moreover, in such cases, it becomes possible to use this technology for faster and more accurate: positioning and guidance of devices onto freight unit and programming certain operating parameters of handling equipment [182]. The energy intensity of LSM that use: radio terminals, docking stations and classic solutions is presented in the rest of the paper (Sect. 6.2).

4.5.2 Electronic Document Interchange in LSM

The essence of EDI has been described in detail by the author in the monographs [297, 299] as well as other publications, such as [144, 145, 158]. The use of EDI in logistic systems allows for paperless, fast interchange of data and electronic documents. This technique allows affecting the energy intensity of warehouse processes, significantly reducing it in some cases: based on the studies on LSM, it was found that in the case of interchange of paper documents between two large companies, 30–50 % of them contained various errors in information. Explaining the resulting complications contributed to extending the trading process by 2–7 days. In addition, about 35 % of the information arrives too late to be able to be taken into account when making various kinds of decisions. In some cases the use of very advanced solutions for handling does not bring about reductions in the energy intensity of moving a palette through the LSM (this case is discussed at the end of the paper, for example, Sect. 6.2.)

Based on the above coefficients, one can calculate the so-called global index, which gives an overview of the situation in the organization with regard to the implementation of EDI. This ratio is calculated as follows:

$$W_{edi} = \sum_{i=1}^{n} w_i \times W_j \tag{4.67}$$

where:
W_{edi}—global index characterizing the usefulness of EDI in the organization (J),
W_i—importance assigned to the i-th partial coefficient,
W_j—value of the i-th coefficient of EDI usefulness.

In Table 4.2 shows coefficients (starting from the most significant), and each of them is assigned significance. The significance of coefficients is the result of the group work of experts dealing with EDI [9].

Based on the Eq. (4.67) the indicators of electronic data interchange are calculated (the higher the value of an indicator, the less energy-intensive the EDI is. The index ranges within W_{edi} = (0 –f · 1) (Tables 4.3, 4.4, 4.5, 4.6 and 4.7).

4.6 Energy Consumption Assessment of LSM

The computational model defined in Sects. 4.1–4.5 herein allows a comparison of the results of calculations of energy intensity with research results on real facilities. It gives the possibility of developing “LSM energy intensity maps” (Fig. 4.18), which allows for comparison of energy intensity of alternative solutions in the LSM as well as any energy intensity of moving a freight unit in any LSM (Figs. 4.19, 4.20, 4.21, 4.22, 4.23, 4.24, 4.25 and 4.26).

Course of the braking process with maximum intensity of 0.25.

The assessment of the energy intensity, e.g. Table 4.8, uses the expert method for evaluating the energy intensity, which works the same way as the Fuzzy logic method. On the basis of previously calculated (according to specified working models of the energy consumption of LSM) two sets of data for the year t and the previous reference year t − 1 (see Fig. 4.27). The method of assessment determines the values of energy intensity or its derivatives for a moment in the future.

The method for evaluating the energy intensity of LSM does not use data for time t, because we have none. Evaluation of the necessary parameters uses the knowledge of an expert who evaluates using criteria in the key areas i.e.

Table 4.2 Significance of coefficients of the usefulness of electronic data interchange

Coefficient position wi	Coefficient significance	Name of coefficient
1	0.19	Intensity
2	0.15	Stabilization
3	0.15	Standardization
4	0.14	Urgency
5	0.12	Time consumption
6	0.08	Paperwork required
7	0.07	Documents sent
8	0.06	Errors
9	0.04	Delays

Source [299]

Table 4.3 Data for calculation

P (truck power W)	4500
F1 (N)	2942
V1 (km/h)	5.5
F2 (N)	1013.5
V2 (km/h)	16
Qw (truck in kg)	3382
Qpal (pallet kg)	2000
Qopor(kg)	75
Qcał. of th truck (kg)	5457
Movement resistance Wc	1012.5

Table 4.4 Calculations of forklift acceleration

V (km/h)	V (m/s)	V average (m/s)	Fk (V) (N)	P (V)	delta S (m)	delta sum S (m)	delta t (s)	delta sum t (s)	delta W (J)	delta sum W (J)
0.00	0.00	0.07	2942.00	0.35	0.0026	0.00	0.04	0.04	7.76	7.76
0.50	0.14	0.20	2942.00	0.35	0.0079	0.01	0.04	0.08	23.29	31.06
1.00	0.27	0.34	2942.00	0.35	0.0132	0.02	0.04	0.12	38.82	69.88
1.50	0.41	0.48	2942.00	0.35	0.0185	0.04	0.04	0.15	54.35	124.23
2.00	0.55	0.61	2942.00	0.35	0.0238	0.07	0.04	0.19	69.88	194.11
2.50	0.68	0.75	2942.00	0.35	0.0290	0.10	0.04	0.23	85.41	279.51
3.00	0.82	0.89	2942.00	0.35	0.0343	0.13	0.04	0.27	100.94	380.45

(continued)

Table 4.4 (continued)

V (km/h)	V (m/s)	V average (m/s)	Fk (V) (N)	P (V)	delta S (m)	delta sum S (m)	delta t (s)	delta sum t (s)	delta W (J)	delta sum W (J)
3.50	0.96	1.02	2942.00	0.35	0.0396	0.17	0.04	0.31	116.46	496.91
4.00	1.09	1.16	2942.00	0.35	0.0449	0.21	0.04	0.35	131.99	628.90
4.50	1.23	1.30	2942.00	0.35	0.0501	0.26	0.04	0.39	147.52	776.43
5.00	1.37	1.43	2942.00	0.35	0.0554	0.32	0.04	0.43	163.05	939.47
5.50	1.50	1.57	2942.00	0.35	0.0607	0.38	0.04	0.46	178.58	1118.05
6.00	1.64	1.71	2942.00	0.35	0.0660	0.45	0.04	0.50	194.11	1312.16
6.50	1.78	1.84	2635.20	0.30	0.0847	0.53	0.05	0.55	223.28	1535.43
7.00	1.91	1.98	2440.00	0.26	0.1034	0.63	0.05	0.60	252.41	1787.85
7.50	2.05	2.12	2271.72	0.23	0.1254	0.76	0.06	0.66	284.78	2072.63
8.00	2.19	2.25	2125.16	0.20	0.1510	0.91	0.07	0.73	320.96	2393.59
8.50	2.32	2.39	1996.36	0.18	0.1811	1.09	0.08	0.80	361.64	2755.23
9.00	2.46	2.53	1882.29	0.16	0.2166	1.31	0.09	0.89	407.73	3162.96
9.50	2.60	2.66	1780.54	0.14	0.2586	1.57	0.10	0.99	460.40	3623.36
10.00	2.73	2.80	1689.23	0.12	0.3085	1.88	0.11	1.10	521.15	4144.50
10.50	2.87	2.94	1606.83	0.11	0.3684	2.24	0.13	1.22	591.99	4736.49
11.00	3.01	3.07	1532.09	0.10	0.4410	2.68	0.14	1.36	675.67	5412.16
11.50	3.14	3.21	1464.00	0.08	0.5301	3.21	0.17	1.53	776.04	6188.20
12.00	3.28	3.35	1401.70	0.07	0.6411	3.86	0.19	1.72	898.62	7086.83
12.50	3.42	3.48	1344.49	0.06	0.7823	4.64	0.22	1.95	1051.73	8138.56
13.00	3.55	3.62	1291.76	0.05	0.9664	5.60	0.27	2.21	1248.37	9386.93
13.50	3.69	3.76	1243.02	0.04	1.2149	6.82	0.32	2.54	1510.20	10897.14
14.00	3.83	3.89	1197.82	0.03	1.5662	8.39	0.40	2.94	1876.07	12773.21
14.50	3.96	4.03	1155.79	0.03	2.0967	10.48	0.52	3.46	2423.36	15196.57
15.00	4.10	4.17	1116.61	0.02	2.9836	13.47	0.72	4.17	3331.50	18528.07
15.50	4.23	4.30	1080.00	0.01	4.7527	18.22	1.10	5.28	5132.89	23660.97
16.00	4.37	4.37	1045.71	0.01	–0.1138	18.11	–0.03	5.25	–119.00	23541.97

Table 4.5 Data for braking calculations

P (forklift power W)	4500
F1 (N)	2942
V1 (km/h)	5.5
F2 (N)	1013.5
V2 (km/h)	16
Qw (forklift in kg)	3382
Qpal (pallet kg)	2000
Qopor (kg)	75
Qcał. Forklift (kg)	5457
Movement resistance Wc	1012.5
Phmax (N)	1364.25
F ham (N)	2376.75

Table 4.6 Calculation results for braking

V (km/h)	V (m/s)	V average (m/s)	Fk (V) (N)	p (V)	delta S (m)	delta sum S (m)	delta t (s)	delta sum t (s)	delta W (J)	delta sum W (J)
0.00	0.00	0.07	2376.75	0.25	0.0037	0.00	0.05	0.05	2372.97	2372.97
0.50	0.14	0.20	2376.75	0.25	0.0112	0.01	0.05	0.11	2365.41	4738.38
1.00	0.27	0.34	2376.75	0.25	0.0187	0.03	0.05	0.16	2357.85	7096.24
1.50	0.41	0.48	2376.75	0.25	0.0261	0.06	0.05	0.22	2350.30	9446.53
2.00	0.55	0.61	2376.75	0.25	0.0336	0.09	0.05	0.27	2342.74	11,789.27
2.50	0.68	0.75	2376.75	0.25	0.0411	0.13	0.05	0.33	2335.18	14,124.45
3.00	0.82	0.89	2376.75	0.25	0.0485	0.18	0.05	0.38	2327.62	16,452.07
3.50	0.96	1.02	2376.75	0.25	0.0560	0.24	0.05	0.44	2320.06	18,772.13
4.00	1.09	1.16	2376.75	0.25	0.0635	0.30	0.05	0.49	2312.50	21,084.63
4.50	1.23	1.30	2376.75	0.25	0.0709	0.37	0.05	0.55	2304.94	23,389.58
5.00	1.37	1.43	2376.75	0.25	0.0784	0.45	0.05	0.60	2297.39	25,686.96
5.50	1.50	1.57	2376.75	0.25	0.0858	0.54	0.05	0.66	2289.83	27,976.79
6.00	1.64	1.71	2376.75	0.25	0.0933	0.63	0.05	0.71	2282.27	30,259.06
6.50	1.78	1.84	2376.75	0.25	0.1008	0.73	0.05	0.77	2274.71	32,533.77
7.00	1.91	1.98	2376.75	0.25	0.1082	0.84	0.05	0.82	2267.15	34,800.92
7.50	2.05	2.12	2376.75	0.25	0.1157	0.96	0.05	0.87	2259.59	37,060.52
8.00	2.19	2.25	2376.75	0.25	0.1232	1.08	0.05	0.93	2252.04	39,312.55
8.50	2.32	2.39	2376.75	0.25	0.1306	1.21	0.05	0.98	2244.48	41,557.03
9.00	2.46	2.53	2376.75	0.25	0.1381	1.35	0.05	1.04	2236.92	43,793.95
9.50	2.60	2.66	2376.75	0.25	0.1456	1.49	0.05	1.09	2229.36	46,023.31
10.00	2.73	2.80	2376.75	0.25	0.1530	1.65	0.05	1.15	2221.80	48,245.11
10.50	2.87	2.94	2376.75	0.25	0.1605	1.81	0.05	1.20	2214.24	50,459.35
11.00	3.01	3.07	2376.75	0.25	0.1680	1.97	0.05	1.26	2206.68	52,666.04
11.50	3.14	3.21	2376.75	0.25	0.1754	2.15	0.05	1.31	2199.13	54,865.17
12.00	3.28	3.35	2376.75	0.25	0.1829	2.33	0.05	1.37	2191.57	57,056.73
12.50	3.42	3.48	2376.75	0.25	0.1904	2.52	0.05	1.42	2184.01	59,240.74
13.00	3.55	3.62	2376.75	0.25	0.1978	2.72	0.05	1.48	2176.45	61,417.19
13.50	3.69	3.76	2376.75	0.25	0.2053	2.93	0.05	1.53	2168.89	63,586.09
14.00	3.83	3.89	2376.75	0.25	0.2128	3.14	0.05	1.58	2161.33	65,747.42
14.50	3.96	4.03	2376.75	0.25	0.2202	3.36	0.05	1.64	2153.78	67,901.20
15.00	4.10	4.17	2376.75	0.25	0.2277	3.59	0.05	1.69	2146.22	70,047.41
15.50	4.23	4.30	2376.75	0.25	0.2352	3.82	0.05	1.75	2138.66	72,186.07
16.00	4.37	4.37	2376.75	0.25	−0.0028	3.82	0.00	1.75	2379.56	74,565.63

Table 4.7 Calculations for minimum-time travel (S1 = 10 m, S2 = 50 m, S3 = 8 m, S4 = 60 m)

V (km/h)	V (m/s)	V average (m/s)	Fk(V) (N)	P (V)	delta S (m)	delta sum S (m)	delta t (s)	delta sum t (s)	delta W (J)	delta sum W (J)
0.00	0.00	0.07	2942.00	0.35	0.0026	0.00	0.04	0.04	7.76	7.76
0.50	0.14	0.20	2942.00	0.35	0.0079	0.01	0.04	0.08	23.29	31.06
1.00	0.27	0.34	2942.00	0.35	0.0132	0.02	0.04	0.12	38.82	69.88
1.50	0.41	0.48	2942.00	0.35	0.0185	0.04	0.04	0.15	54.35	124.23
2.00	0.55	0.61	2942.00	0.35	0.0238	0.07	0.04	0.19	69.88	194.11
2.50	0.68	0.75	2942.00	0.35	0.0290	0.10	0.04	0.23	85.41	279.51
3.00	0.82	0.89	2942.00	0.35	0.0343	0.13	0.04	0.27	100.94	380.45
3.50	0.96	1.02	2942.00	0.35	0.0396	0.17	0.04	0.31	116.46	496.91
4.00	1.09	1.16	2942.00	0.35	0.0449	0.21	0.04	0.35	131.99	628.90
4.50	1.23	1.30	2942.00	0.35	0.0501	0.26	0.04	0.39	147.52	776.43
5.00	1.37	1.43	2942.00	0.35	0.0554	0.32	0.04	0.43	163.05	939.47
5.50	1.50	1.57	2942.00	0.35	0.0607	0.38	0.04	0.46	178.58	1118.05
6.00	1.64	1.71	2942.00	0.35	0.0660	0.45	0.04	0.50	194.11	1312.16
6.50	1.78	1.84	2635.20	0.30	0.0847	0.53	0.05	0.55	223.28	1535.43
7.00	1.91	1.98	2440.00	0.26	0.1034	0.63	0.05	0.60	252.41	1787.85
7.50	2.05	2.12	2271.72	0.23	0.1254	0.76	0.06	0.66	284.78	2072.63
8.00	2.19	2.25	2125.16	0.20	0.1510	0.91	0.07	0.73	320.96	2393.59
8.50	2.32	2.39	1996.36	0.18	0.1811	1.09	0.08	0.80	361.64	2755.23
9.00	2.46	2.53	1882.29	0.16	0.2166	1.31	0.09	0.89	407.73	3162.96
9.50	2.60	2.66	1780.54	0.14	0.2586	1.57	0.10	0.99	460.40	3623.36
10.00	2.73	2.80	1689.23	0.12	0.3085	1.88	0.11	1.10	521.15	4144.50
10.50	2.87	2.94	1606.83	0.11	0.3684	2.24	0.13	1.22	591.99	4736.49
11.00	3.01	3.07	1532.09	0.10	0.4410	2.68	0.14	1.36	675.67	5412.16
11.50	3.14	3.21	1464.00	0.08	0.5301	3.21	0.17	1.53	776.04	6188.20
12.00	3.28	3.35	1401.70	0.07	0.6411	3.86	0.19	1.72	898.62	7086.83
12.50	3.42	3.48	1344.49	0.06	0.7823	4.64	0.22	1.95	1051.73	8138.56
13.00	3.55	3.48	1291.76	0.05	0.9299	5.57	0.27	2.21	1201.27	9339.82
12.50	3.42	3.35	1291.76	0.05	0.8935	6.46	0.27	2.48	1154.16	10,493.98
12.00	3.28	3.21	2376.75	0.25	0.1754	6.64	0.05	2.53	416.96	10,910.94
11.50	3.14	3.07	2376.75	0.25	0.1680	6.81	0.05	2.59	399.21	11,310.15
11.00	3.01	2.94	2376.75	0.25	0.1605	6.97	0.05	2.64	381.47	11,691.62
10.50	2.87	2.80	2376.75	0.25	0.1530	7.12	0.05	2.70	363.73	12,055.35
10.00	2.73	2.66	2376.75	0.25	0.1456	7.26	0.05	2.75	345.98	12,401.33
9.50	2.60	2.53	2376.75	0.25	0.1381	7.40	0.05	2.81	328.24	12,729.57
9.00	2.46	2.39	2376.75	0.25	0.1306	7.53	0.05	2.86	310.50	13,040.07
8.50	2.32	2.25	2376.75	0.25	0.1232	7.66	0.05	2.92	292.76	13,332.83
8.00	2.19	2.12	2376.75	0.25	0.1157	7.77	0.05	2.97	275.01	13,607.84
7.50	2.05	1.98	2376.75	0.25	0.1082	7.88	0.05	3.03	257.27	13,865.11
7.00	1.91	1.84	2376.75	0.25	0.1008	7.98	0.05	3.08	239.53	14,104.64
6.50	1.78	1.71	2376.75	0.25	0.0933	8.07	0.05	3.14	221.78	14,326.42
6.00	1.64	1.57	2376.75	0.25	0.0858	8.16	0.05	3.19	204.04	14,530.46
5.50	1.50	1.43	2376.75	0.25	0.0784	8.24	0.05	3.24	186.30	14,716.76

(continued)

Table 4.7 (continued)

V (km/h)	V (m/s)	V average (m/s)	Fk(V) (N)	P (V)	delta S (m)	delta sum S (m)	delta t (s)	delta sum t (s)	delta W (J)	delta sum W (J)
5.00	1.37	1.30	2376.75	0.25	0.0709	8.31	0.05	3.30	168.56	14,885.32
4.50	1.23	1.16	2376.75	0.25	0.0635	8.37	0.05	3.35	150.81	15,036.13
4.00	1.09	1.02	2376.75	0.25	0.0560	8.43	0.05	3.41	133.07	15,169.20
3.50	0.96	0.89	2376.75	0.25	0.0485	8.48	0.05	3.46	115.33	15,284.53
3.00	0.82	0.75	2376.75	0.25	0.0411	8.52	0.05	3.52	97.59	15,382.11
2.50	0.68	0.61	2376.75	0.25	0.0336	8.55	0.05	3.57	79.84	15,461.96
2.00	0.55	0.48	2376.75	0.25	0.0261	8.58	0.05	3.63	62.10	15,524.06
1.50	0.41	0.34	2376.75	0.25	0.0187	8.60	0.05	3.68	44.36	15,568.41
1.00	0.27	0.20	2376.75	0.25	0.0112	8.61	0.05	3.74	26.61	15,595.03
0.50	0.14	0.07	2376.75	0.25	0.0037	8.61	0.05	3.79	8.87	15,603.90
0.00	0.00	0.07	2376.75	0.25	−0.0037	8.61	−0.05	3.74	−8.87	15,595.03
0.50	0.14	0.20	2942.00	0.35	0.0079	8.62	0.04	3.77	23.29	15,618.32
1.00	0.27	0.34	2942.00	0.35	0.0132	8.63	0.04	3.81	38.82	15,657.14
1.50	0.41	0.48	2942.00	0.35	0.0185	8.65	0.04	3.85	54.35	15,711.49
2.00	0.55	0.61	2942.00	0.35	0.0238	8.67	0.04	3.89	69.88	15,781.37
2.50	0.68	0.75	2942.00	0.35	0.0290	8.70	0.04	3.93	85.41	15,866.78
3.00	0.82	0.89	2942.00	0.35	0.0343	8.73	0.04	3.97	100.94	15,967.71
3.50	0.96	1.02	2942.00	0.35	0.0396	8.77	0.04	4.01	116.46	16,084.18
4.00	1.09	1.16	2942.00	0.35	0.0449	8.82	0.04	4.05	131.99	16,216.17
4.50	1.23	1.30	2942.00	0.35	0.0501	8.87	0.04	4.08	147.52	16,363.69
5.00	1.37	1.43	2942.00	0.35	0.0554	8.92	0.04	4.12	163.05	16,526.74
5.50	1.50	1.57	2942.00	0.35	0.0607	8.99	0.04	4.16	178.58	16,705.32
6.00	1.64	1.71	2942.00	0.35	0.0660	9.05	0.04	4.20	194.11	16,899.42
6.50	1.78	1.84	2635.20	0.30	0.0847	9.14	0.05	4.25	223.28	17,122.70
7.00	1.91	1.98	2440.00	0.26	0.1034	9.24	0.05	4.30	252.41	17,375.11
7.50	2.05	2.12	2271.72	0.23	0.1254	9.36	0.06	4.36	284.78	17,659.90
8.00	2.19	2.25	2125.16	0.20	0.1510	9.52	0.07	4.42	320.96	17,980.85
8.50	2.32	2.39	1996.36	0.18	0.1811	9.70	0.08	4.50	361.64	18,342.49
9.00	2.46	2.53	1882.29	0.16	0.2166	9.91	0.09	4.59	407.73	18,750.22
9.50	2.60	2.66	1780.54	0.14	0.2586	10.17	0.10	4.68	460.40	19,210.62
10.00	2.73	2.80	1689.23	0.12	0.3085	10.48	0.11	4.79	521.15	19,731.77
10.50	2.87	2.94	1606.83	0.11	0.3684	10.85	0.13	4.92	591.99	20,323.76
11.00	3.01	3.07	1532.09	0.10	0.4410	11.29	0.14	5.06	675.67	20,999.43
11.50	3.14	3.21	1464.00	0.08	0.5301	11.82	0.17	5.23	776.04	21,775.47
12.00	3.28	3.35	1401.70	0.07	0.6411	12.46	0.19	5.42	898.62	22,674.09
12.50	3.42	3.48	1344.49	0.06	0.7823	13.24	0.22	5.64	1051.73	23,725.82
13.00	3.55	3.62	1291.76	0.05	0.9664	14.21	0.27	5.91	1248.37	24,974.20
13.50	3.69	3.76	1243.02	0.04	1.2149	15.42	0.32	6.23	1510.20	26,484.40
14.00	3.83	3.89	1197.82	0.03	1.5662	16.99	0.40	6.64	1876.07	28,360.47
14.50	3.96	4.03	1155.79	0.03	2.0967	19.09	0.52	7.16	2423.36	30,,783.84
15.00	4.10	4.17	1116.61	0.02	2.9836	22.07	0.72	7.87	3331.50	34115.34

(continued)

Table 4.7 (continued)

V (km/h)	V (m/s)	V average (m/s)	Fk(V) (N)	P (V)	delta S (m)	delta sum S (m)	delta t (s)	delta sum t (s)	delta W (J)	delta sum W (J)
15.50	4.23	4.17	1012.50	0.00	41.4300	63.50	9.94	17.82	41947.88	76,063.21
15.00	4.10	4.03	2376.75	0.25	0.2202	63.72	0.05	17.87	523.41	76,586.62
14.50	3.96	3.89	2376.75	0.25	0.2128	63.93	0.05	17.92	505.67	77,092.29
14.00	3.83	3.76	2376.75	0.25	0.2053	64.14	0.05	17.98	487.93	77,580.22
13.50	3.69	3.62	2376.75	0.25	0.1978	64.34	0.05	18.03	470.18	78,050.40
13.00	3.55	3.48	2376.75	0.25	0.1904	64.53	0.05	18.09	452.44	78,502.84
12.50	3.42	3.35	2376.75	0.25	0.1829	64.71	0.05	18.14	434.70	78,937.54
12.00	3.28	3.21	2376.75	0.25	0.1754	64.89	0.05	18.20	416.96	79,354.50
11.50	3.14	3.07	2376.75	0.25	0.1680	65.05	0.05	18.25	399.21	79,753.71
11.00	3.01	2.94	2376.75	0.25	0.1605	65.21	0.05	18.31	381.47	80,135.18
10.50	2.87	2.80	2376.75	0.25	0.1530	65.37	0.05	18.36	363.73	80,498.90
10.00	2.73	2.66	2376.75	0.25	0.1456	65.51	0.05	18.42	345.98	80,844.89
9.50	2.60	2.53	2376.75	0.25	0.1381	65.65	0.05	18.47	328.24	81,173.13
9.00	2.46	2.39	2376.75	0.25	0.1306	65.78	0.05	18.53	310.50	81,483.63
8.50	2.32	2.25	2376.75	0.25	0.1232	65.91	0.05	18.58	292.76	81,776.38
8.00	2.19	2.12	2376.75	0.25	0.1157	66.02	0.05	18.63	275.01	82,051.40
7.50	2.05	1.98	2376.75	0.25	0.1082	66.13	0.05	18.69	257.27	82,308.67
7.00	1.91	1.84	2376.75	0.25	0.1008	66.23	0.05	18.74	239.53	82,548.19
6.50	1.78	1.71	2376.75	0.25	0.0933	66.32	0.05	18.80	221.78	82,769.98
6.00	1.64	1.57	2376.75	0.25	0.0858	66.41	0.05	18.85	204.04	82,974.02
5.50	1.50	1.43	2376.75	0.25	0.0784	66.49	0.05	18.91	186.30	83,160.32
5.00	1.37	1.30	2376.75	0.25	0.0709	66.56	0.05	18.96	168.56	83,328.87
4.50	1.23	1.16	2376.75	0.25	0.0635	66.62	0.05	19.02	150.81	83,479.69
4.00	1.09	1.02	2376.75	0.25	0.0560	66.68	0.05	19.07	133.07	83,612.76
3.50	0.96	0.89	2376.75	0.25	0.0485	66.73	0.05	19.13	115.33	83,728.09
3.00	0.82	0.75	2376.75	0.25	0.0411	66.77	0.05	19.18	97.59	83,825.67
2.50	0.68	0.61	2376.75	0.25	0.0336	66.80	0.05	19.24	79.84	83,905.51
2.00	0.55	0.48	2376.75	0.25	0.0261	66.83	0.05	19.29	62.10	83,967.61
1.50	0.41	0.34	2376.75	0.25	0.0187	66.85	0.05	19.35	44.36	84,011.97
1.00	0.27	0.20	2376.75	0.25	0.0112	66.86	0.05	19.40	26.61	84,038.59
0.50	0.14	0.07	2376.75	0.25	0.0037	66.86	0.05	19.45	8.87	84,047.46
0.00	0.00	0.07	2376.75	0.25	–0.0037	66.86	–0.05	19.40	−8.87	84,038.59
0.50	0.14	0.20	2942.00	0.35	0.0079	66.86	0.04	19.44	23.29	84,061.88
1.00	0.27	0.34	2942.00	0.35	0.0132	66.88	0.04	19.48	38.82	84,100.70
1.50	0.41	0.48	2942.00	0.35	0.0185	66.90	0.04	19.52	54.35	84,155.05
2.00	0.55	0.61	2942.00	0.35	0.0238	66.92	0.04	19.55	69.88	84,224.93
2.50	0.68	0.75	2942.00	0.35	0.0290	66.95	0.04	19.59	85.41	84,310.33
3.00	0.82	0.89	2942.00	0.35	0.0343	66.98	0.04	19.63	100.94	84,411.27
3.50	0.96	1.02	2942.00	0.35	0.0396	67.02	0.04	19.67	116.46	84,527.73
4.00	1.09	1.16	2942.00	0.35	0.0449	67.07	0.04	19.71	131.99	84,659.73
4.50	1.23	1.30	2942.00	0.35	0.0501	67.12	0.04	19.75	147.52	84,807.25

(continued)

Table 4.7 (continued)

V (km/h)	V (m/s)	V average (m/s)	Fk(V) (N)	P (V)	delta S (m)	delta sum S (m)	delta t (s)	delta sum t (s)	delta W (J)	delta sum W (J)
5.00	1.37	1.43	2942.00	0.35	0.0554	67.17	0.04	19.79	163.05	84,970.30
5.50	1.50	1.57	2942.00	0.35	0.0607	67.23	0.04	19.82	178.58	85,148.87
6.00	1.64	1.71	2942.00	0.35	0.0660	67.30	0.04	19.86	194.11	85,342.98
6.50	1.78	1.84	2635.20	0.30	0.0847	67.39	0.05	19.91	223.28	85,566.26
7.00	1.91	1.98	2440.00	0.26	0.1034	67.49	0.05	19.96	252.41	85,818.67
7.50	2.05	2.12	2271.72	0.23	0.1254	67.61	0.06	20.02	284.78	86,103.45
8.00	2.19	2.25	2125.16	0.20	0.1510	67.76	0.07	20.09	320.96	86,424.41
8.50	2.32	2.39	1996.36	0.18	0.1811	67.95	0.08	20.16	361.64	86,786.05
9.00	2.46	2.53	1882.29	0.16	0.2166	68.16	0.09	20.25	407.73	87,193.78
9.50	2.60	2.66	1780.54	0.14	0.2586	68.42	0.10	20.35	460.40	87,654.18
10.00	2.73	2.80	1689.23	0.12	0.3085	68.73	0.11	20.46	521.15	88,175.33
10.50	2.87	2.94	1606.83	0.11	0.3684	69.10	0.13	20.58	591.99	88,767.31
11.00	3.01	3.07	1532.09	0.10	0.4410	69.54	0.14	20.73	675.67	89,442.99
11.50	3.14	3.21	1464.00	0.08	0.5301	70.07	0.17	20.89	776.04	90,219.02
12.00	3.28	3.35	1401.70	0.07	0.6411	70.71	0.19	21.08	898.62	91,117.65
12.50	3.42	3.48	1344.49	0.06	0.7823	71.49	0.22	21.31	1051.73	92,169.38
13.00	3.55	3.62	1291.76	0.05	0.9664	72.46	0.27	21.57	1248.37	93,417.75
13.50	3.69	3.76	1243.02	0.04	1.2149	73.67	0.32	21.90	1510.20	94,927.96
14.00	3.83	3.76	1197.82	0.03	–1.5113	72.16	–0.40	21.49	–1810.24	93,117.71
13.50	3.69	3.62	2376.75	0.25	0.1978	72.36	0.05	21.55	470.18	93,587.90
13.00	3.55	3.48	2376.75	0.25	0.1904	72.55	0.05	21.60	452.44	94,040.34
12.50	3.42	3.35	2376.75	0.25	0.1829	72.73	0.05	21.66	434.70	94,475.04
12.00	3.28	3.21	2376.75	0.25	0.1754	72.91	0.05	21.71	416.96	94,891.99
11.50	3.14	3.07	2376.75	0.25	0.1680	73.08	0.05	21.77	399.21	95,291.20
11.00	3.01	2.94	2376.75	0.25	0.1605	73.24	0.05	21.82	381.47	95,672.67
10.50	2.87	2.80	2376.75	0.25	0.1530	73.39	0.05	21.88	363.73	96,036.40
10.00	2.73	2.66	2376.75	0.25	0.1456	73.54	0.05	21.93	345.98	96,382.38
9.50	2.60	2.53	2376.75	0.25	0.1381	73.67	0.05	21.99	328.24	96,710.62
9.00	2.46	2.39	2376.75	0.25	0.1306	73.81	0.05	22.04	310.50	97,021.12
8.50	2.32	2.25	2376.75	0.25	0.1232	73.93	0.05	22.10	292.76	97,313.88
8.00	2.19	2.12	2376.75	0.25	0.1157	74.04	0.05	22.15	275.01	97,588.89
7.50	2.05	1.98	2376.75	0.25	0.1082	74.15	0.05	22.21	257.27	97,846.16
7.00	1.91	1.84	2376.75	0.25	0.1008	74.25	0.05	22.26	239.53	98,085.69
6.50	1.78	1.71	2376.75	0.25	0.0933	74.35	0.05	22.31	221.78	98,307.47
6.00	1.64	1.57	2376.75	0.25	0.0858	74.43	0.05	22.37	204.04	98,511.51
5.50	1.50	1.43	2376.75	0.25	0.0784	74.51	0.05	22.42	186.30	98,697.81
5.00	1.37	1.30	2376.75	0.25	0.0709	74.58	0.05	22.48	168.56	98,866.37
4.50	1.23	1.16	2376.75	0.25	0.0635	74.64	0.05	22.53	150.81	99,017.18
4.00	1.09	1.02	2376.75	0.25	0.0560	74.70	0.05	22.59	133.07	99,150.25
3.50	0.96	0.89	2376.75	0.25	0.0485	74.75	0.05	22.64	115.33	99,265.58
3.00	0.82	0.75	2376.75	0.25	0.0411	74.79	0.05	22.70	97.59	99,363.17

(continued)

Table 4.7 (continued)

V (km/h)	V (m/s)	V average (m/s)	Fk(V) (N)	P (V)	delta S (m)	delta sum S (m)	delta t (s)	delta sum t (s)	delta W (J)	delta sum W (J)
2.50	0.68	0.61	2376.75	0.25	0.0336	74.82	0.05	22.75	79.84	99,443.01
2.00	0.55	0.48	2376.75	0.25	0.0261	74.85	0.05	22.81	62.10	99,505.11
1.50	0.41	0.34	2376.75	0.25	0.0187	74.87	0.05	22.86	44.36	99,549.47
1.00	0.27	0.20	2376.75	0.25	0.0112	74.88	0.05	22.92	26.61	99,576.08
0.50	0.14	0.07	2376.75	0.25	0.0037	74.88	0.05	22.97	8.87	99,584.95
0.00	0.00	0.07	2376.75	0.25	−0.0037	74.88	−0.05	22.92	−8.87	99,576.08
0.50	0.14	0.20	2942.00	0.35	0.0079	74.89	0.04	22.95	23.29	99,599.37
1.00	0.27	0.34	2942.00	0.35	0.0132	74.90	0.04	22.99	38.82	99,638.19
1.50	0.41	0.48	2942.00	0.35	0.0185	74.92	0.04	23.03	54.35	99,692.54
2.00	0.55	0.61	2942.00	0.35	0.0238	74.94	0.04	23.07	69.88	99,762.42
2.50	0.68	0.75	2942.00	0.35	0.0290	74.97	0.04	23.11	85.41	99,847.83
3.00	0.82	0.89	2942.00	0.35	0.0343	75.01	0.04	23.15	100.94	99,948.76
3.50	0.96	1.02	2942.00	0.35	0.0396	75.05	0.04	23.19	116.46	100,065.23
4.00	1.09	1.16	2942.00	0.35	0.0449	75.09	0.04	23.22	131.99	100,197.22
4.50	1.23	1.30	2942.00	0.35	0.0501	75.14	0.04	23.26	147.52	100,344.74
5.00	1.37	1.43	2942.00	0.35	0.0554	75.20	0.04	23.30	163.05	100,507.79
5.50	1.50	1.57	2942.00	0.35	0.0607	75.26	0.04	23.34	178.58	100,686.37
6.00	1.64	1.71	2942.00	0.35	0.0660	75.32	0.04	23.38	194.11	100,880.48
6.50	1.78	1.84	2635.20	0.30	0.0847	75.41	0.05	23.43	223.28	101,103.75
7.00	1.91	1.98	2440.00	0.26	0.1034	75.51	0.05	23.48	252.41	101,356.16
7.50	2.05	2.12	2271.72	0.23	0.1254	75.64	0.06	23.54	284.78	101,640.95
8.00	2.19	2.25	2125.16	0.20	0.1510	75.79	0.07	23.60	320.96	101,961.90
8.50	2.32	2.39	1996.36	0.18	0.1811	75.97	0.08	23.68	361.64	102,323.54
9.00	2.46	2.53	1882.29	0.16	0.2166	76.19	0.09	23.77	407.73	102,731.28
9.50	2.60	2.66	1780.54	0.14	0.2586	76.44	0.10	23.86	460.40	103,191.68
10.00	2.73	2.80	1689.23	0.12	0.3085	76.75	0.11	23.97	521.15	103,712.82
10.50	2.87	2.94	1606.83	0.11	0.3684	77.12	0.13	24.10	591.99	104,304.81
11.00	3.01	3.07	1532.09	0.10	0.4410	77.56	0.14	24.24	675.67	104,980.48
11.50	3.14	3.21	1464.00	0.08	0.5301	78.09	0.17	24.41	776.04	105,756.52
12.00	3.28	3.35	1401.70	0.07	0.6411	78.73	0.19	24.60	898.62	106,655.14
12.50	3.42	3.48	1344.49	0.06	0.7823	79.52	0.22	24.82	1051.73	107,706.87
13.00	3.55	3.62	1291.76	0.05	0.9664	80.48	0.27	25.09	1248.37	108,955.25
13.50	3.69	3.76	1243.02	0.04	1.2149	81.70	0.32	25.41	1510.20	110,465.45
14.00	3.83	3.89	1197.82	0.03	1.5662	83.26	0.40	25.82	1876.07	112,341.53
14.50	3.96	4.03	1155.79	0.03	2.0967	85.36	0.52	26.34	2423.36	114,764.89
15.00	4.10	4.17	1116.61	0.02	2.9836	88.34	0.72	27.05	3331.50	118,096.39
15.50	4.23	4.17	1012.50	0.00	51.4300	139.77	12.34	39.39	52072.88	170,169.26
15.00	4.10	4.03	2376.75	0.25	0.2202	139.99	0.05	39.45	523.41	170,692.68
14.50	3.96	3.89	2376.75	0.25	0.2128	140.21	0.05	39.50	505.67	171,198.35
14.00	3.83	3.76	2376.75	0.25	0.2053	140.41	0.05	39.56	487.93	171,686.27
13.50	3.69	3.62	2376.75	0.25	0.1978	140.61	0.05	39.61	470.18	172,156.45

(continued)

Table 4.7 (continued)

V (km/h)	V (m/s)	V average (m/s)	Fk(V) (N)	P (V)	delta S (m)	delta sum S (m)	delta t (s)	delta sum t (s)	delta W (J)	delta sum W (J)
13.00	3.55	3.48	2376.75	0.25	0.1904	140.80	0.05	39.67	452.44	172,608.90
12.50	3.42	3.35	2376.75	0.25	0.1829	140.98	0.05	39.72	434.70	173,043.59
12.00	3.28	3.21	2376.75	0.25	0.1754	141.16	0.05	39.78	416.96	173,460.55
11.50	3.14	3.07	2376.75	0.25	0.1680	141.33	0.05	39.83	399.21	173,859.76
11.00	3.01	2.94	2376.75	0.25	0.1605	141.49	0.05	39.89	381.47	174,241.23
10.50	2.87	2.80	2376.75	0.25	0.1530	141.64	0.05	39.94	363.73	174,604.96
10.00	2.73	2.66	2376.75	0.25	0.1456	141.79	0.05	40.00	345.98	174,950.94
9.50	2.60	2.53	2376.75	0.25	0.1381	141.92	0.05	40.05	328.24	175,279.18
9.00	2.46	2.39	2376.75	0.25	0.1306	142.05	0.05	40.10	310.50	175,589.68
8.50	2.32	2.25	2376.75	0.25	0.1232	142.18	0.05	40.16	292.76	175,882.44
8.00	2.19	2.12	2376.75	0.25	0.1157	142.29	0.05	40.21	275.01	176,157.45
7.50	2.05	1.98	2376.75	0.25	0.1082	142.40	0.05	40.27	257.27	176,414.72
7.00	1.91	1.84	2376.75	0.25	0.1008	142.50	0.05	40.32	239.53	176,654.25
6.50	1.78	1.71	2376.75	0.25	0.0933	142.60	0.05	40.38	221.78	176,876.03
6.00	1.64	1.57	2376.75	0.25	0.0858	142.68	0.05	40.43	204.04	177.080.07
5.50	1.50	1.43	2376.75	0.25	0.0784	142.76	0.05	40.49	186.30	177.266.37
5.00	1.37	1.30	2376.75	0.25	0.0709	142.83	0.05	40.54	168.56	177.434.93
4.50	1.23	1.16	2376.75	0.25	0.0635	142.89	0.05	40.60	150.81	177.585.74
4.00	1.09	1.02	2376.75	0.25	0.0560	142.95	0.05	40.65	133.07	177.718.81
3.50	0.96	0.89	2376.75	0.25	0.0485	143.00	0.05	40.71	115.33	177.834.14
3.00	0.82	0.75	2376.75	0.25	0.0411	143.04	0.05	40.76	97.59	177.931.73
2.50	0.68	0.61	2376.75	0.25	0.0336	143.07	0.05	40.82	79.84	178,011.57
2.00	0.55	0.48	2376.75	0.25	0.0261	143.10	0.05	40.87	62.10	178,073.67
1.50	0.41	0.34	2376.75	0.25	0.0187	143.12	0.05	40.92	44.36	178,118.02
1.00	0.27	0.20	2376.75	0.25	0.0112	143.13	0.05	40.98	26.61	178,144.64
0.50	0.14	0.07	2376.75	0.25	0.0037	143.13	0.05	41.03	8.87	178,153.51
0.00	0.00	0.00	2376.75	0.25	0.0000	143.13	#DZIEL/0!	#DZIEL/0!	0.00	178,153.51

technological capabilities, the economy, and takes into account the trends in technical and organizational solutions. The basis of the method is the following formula:

$$\frac{E_{t-1}}{E_t} = \sum_p ec_{p,t} \times \frac{E_{p,t-1}}{E_{p,t}} \tag{4.68}$$

where:
E_t—energy intensity of LSM in the year t,
ecp,t—process share ratio $p_m^1(\theta) \div p_m^4(\theta)$ in energy intensity in LSM in the year t.

This formula states that coefficient $\frac{E_{t-1}}{E_t}$ is the weighted mean of the indicators of process energy intensity $\frac{E_{p,t-1}}{E_{p,t}}$ (for all processes) with significance equal to the share ratio of the process *p* in the energy intensity of LSM in the year t.

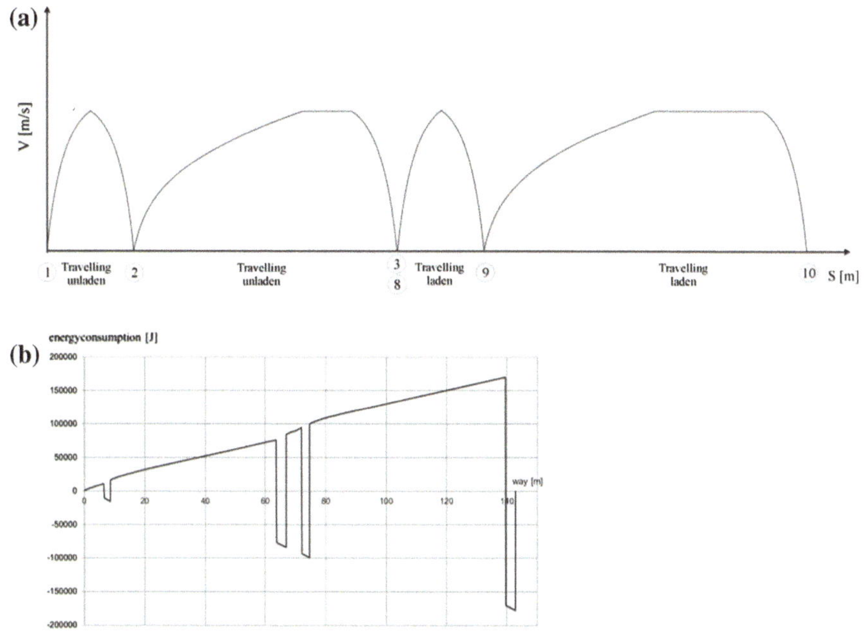

Fig. 4.18 Map of energy intensity of moving a pallet through the warehouse (assumptions as in Fig. 4.9). **a** Diagram of movement; **b** energy consumed (below the axis of energy recovery). Author's own work

Fig. 4.19 Traction characteristics of forklift EFG2-20

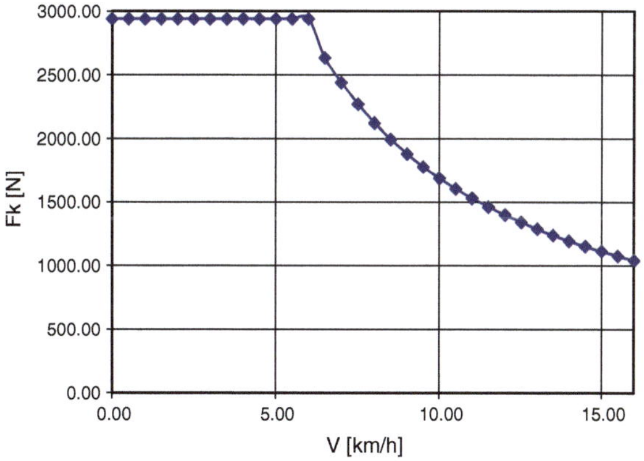

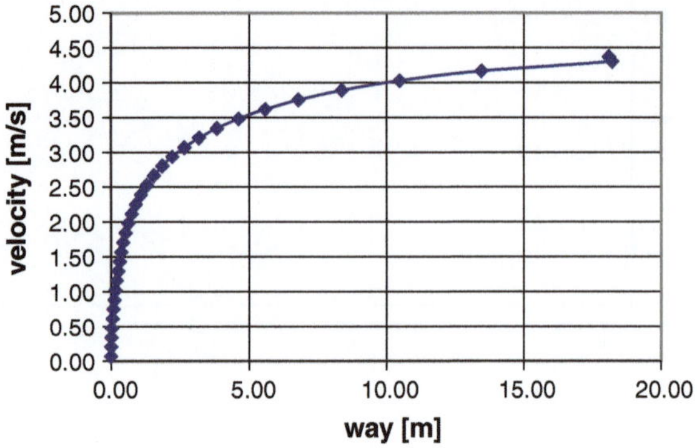

Fig. 4.20 Course of acceleration of EFG-220 forklift with load

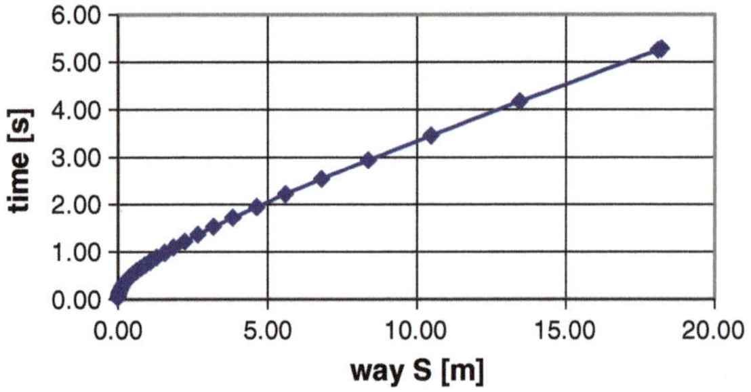

Fig. 4.21 Acceleration time for EFG-220 forklift

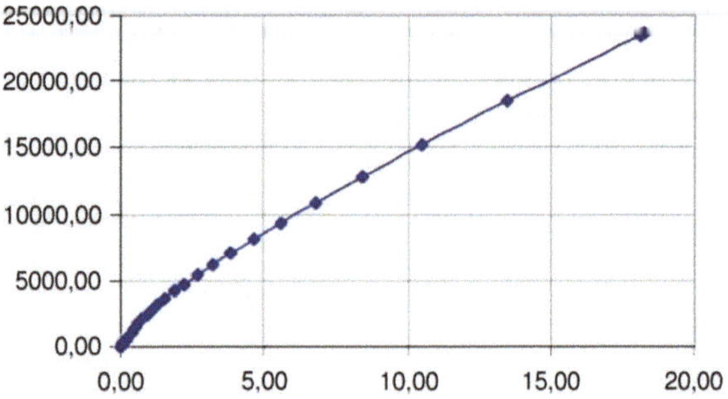

Fig. 4.22 Energy consumption on traction during acceleration of the EFG-220 forklift

With these markings $\frac{E_{t-1}}{E_t}$ is the main index for the whole LSM. The main energy intensity indicator (optional process, technology) is convertible to the following form:

$$\frac{E_{p,t-1}}{E_{p,t}} = 1 + w_{p,t} \cdot \Delta E_{p,t}^{\max} \tag{4.69}$$

where $\Delta E_{p,t}^{max}$ is the maximum possible, the theoretical value of improving the energy intensity p; this value can be estimated based on theoretical sources (model for energy intensity in this paper)

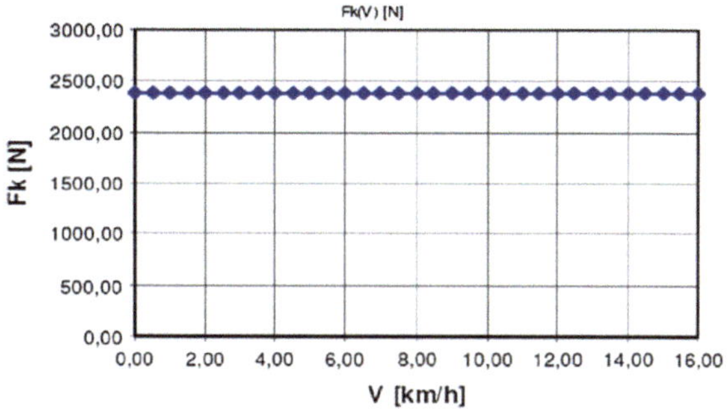

Fig. 4.23 Characteristics of a forklift truck braking system EFG-220

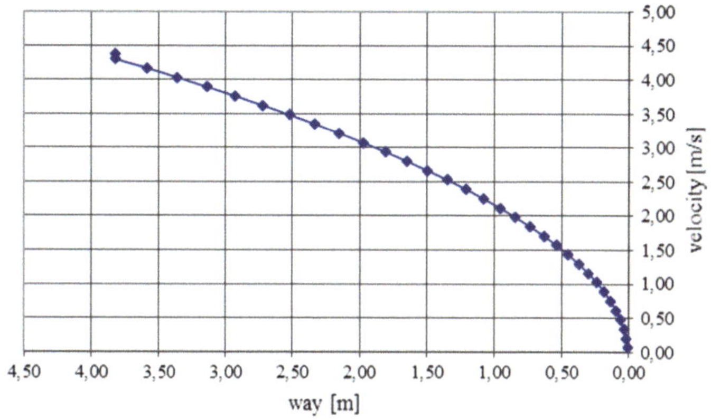

Fig. 4.24 Course of braking speed for forklift EFG-220 with load

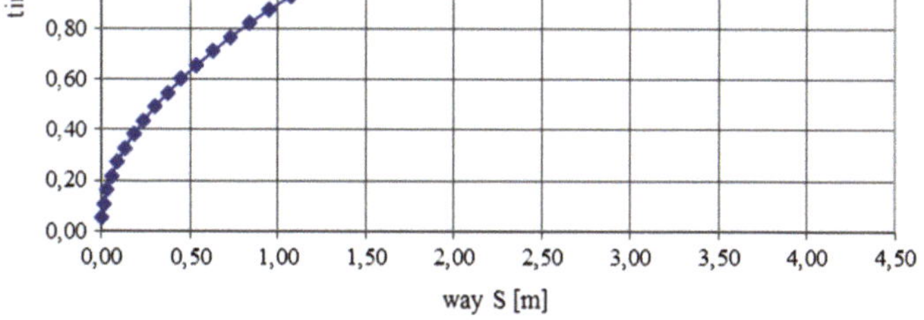

Fig. 4.25 Breaking time for forklift EFG-220

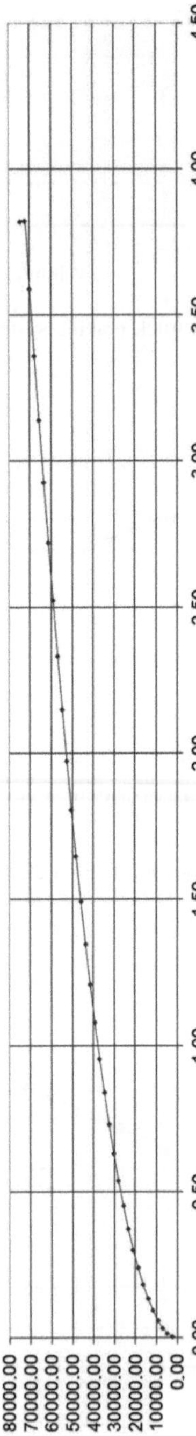

Fig. 4.26 Energy consumption purposes forklift traction when braking

Table 4.8 Results of using the evaluation model of energy consumption

Total energy intensity (J)	17,535,613.92
Braking energy intensity (J)	10,709,790.07
Energy intensity for V-const. (J)	246,232.48

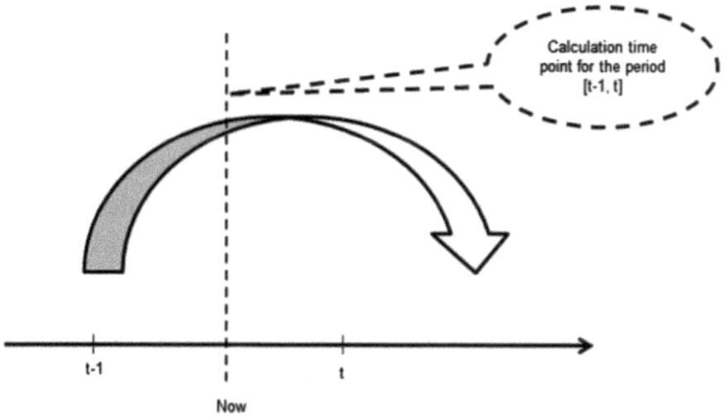

Fig. 4.27 The time points of the methods of energy intensity evaluation. Author's own work

$w_{p,t}$ weighting factor whose value is estimated, and it takes the value from the range [0, 1] and could be interpreted as the probability of maximum effect $\Delta E_{p,t}^{max}$ in the year t.

In many cases—in small warehouses (though not exclusively), where there are only forklifts, and warehouses, where the shortest service time is not important, because the device will be unused after the completion of tasks—for such cases, based on the above considerations, one can select the optimal forklift speed from the point of view of energy consumption [315].

If a forklift with daily consumption $C_{day}(V)$ (kg/h) works 24 h at a constant speed v(km/h), the resulting fuel consumption is given in the formula:

$$CB(V) = C_{day}(V)/24 \cdot V \tag{4.70}$$

On the basis of equations for the movement of the forklift, one can calculate the relationship between speed and daily consumption $C_{day}(v)$. The speed-dependent fuel consumption by auxiliary units, air-conditioning, etc., which has no direct effect on the forklift drive, is to be added to the cost of using the forklift. Using the formula (4.70), e.g. based on the relation between the speed of the forklift with a capacity of 2.5 tons and daily consumption, one can calculated the dependency of speed ratio on fuel consumption required for movement of cargo (vertically and horizontally), as shown in Fig. 4.7. Similar curves for consumption are obtained for other types of forklifts [1, 3, 4]. The curve of fuel consumption by a forklift rises from the minimum speed v_{min}, the value of which amounts to roughly half of the interpretation value, with increasing speed to reach the maximum speed v_{max}. However, depending on the type of forklift, its structure, drive type, degree of loading and other factors, basic consumption will be higher or lower, the speed will vary, the growth rate may vary. This is confirmed by the papers [2, 6, 8].

In the case of calculations aimed at optimization, it seems most sensible to provide a wear curve by approximation (4.71):

$$C_B(V) = C_0 + C_1 \cdot V^n \tag{4.71}$$

Using the method of least squares, both wear parameters C0 and C1 and the velocity exponent n can be defined in such a way that the sum of the deviations of the measured values of the squared approximation is minimal [7]. With a total of three parameters, at least four measured values must be known. In the case of this forklift, Fig. 4.7 shows how well the measured mileage can be represented using approximation (4.71) with the following values C0 = 58, C1 = 0.0013 and n = 4.5. For other types of forklifts (e.g. towing forklifts) a similar good approximation yields parameter values that deviate significantly. The speed exponents are from n $\approx$ 3 for small forklifts to n n $\approx$ 7 for large trucks. This means that fuel consumption per meter of cargo movement by the forklift increases with the increasing speed to the power of three to seven.

In Fig. 4.27 it can be seen that in the case of an exemplary forklift, fuel consumption can be reduced by about a half by reducing its speed from the maximum speed of 25 km/h by approx. 20 % to 20 km/h. Slowing down to the speed that is optimal from the point of view of fuel consumption, which is achieved at the minimum speed of 12.5 km/h gives a reduction in fuel consumption of approx. 75 %.

The travel time of a forklift truck on the path with a total length of $\mathrm{L} = \sum_{\mathrm{i}=1}^{\mathrm{N_s}} \mathrm{L_i}$, divided into i = 1 sections *L*, *i* = 1,2,…,*Ns*, which are traversed at different speeds s V, is calculated based on the following formula (4.3):

$$T_F = \sum \frac{L_i}{V_i} = \frac{L}{V_m}$$

The average speed is calculated according to the formula (4.69):

$$V_m = \frac{L}{\left(\sum \frac{Li}{Vi}\right)}$$

with this type of driving, fuel consumption is calculated according to the formula (4.70):

$$C_B(L) = \sum C_B(V_i) \cdot L_i = \sum \left(C_0 + C_1 \cdot V_i^n\right) \cdot L_i$$

If the speed on particular sections differ from the average speed by Ai, after inserting $V_t = V_m + 4$ to the Eq. (4.70), fuel consumption will be higher than when the forklift had traveled all sections at a one constant average speed while maintaining the same total driving time. This applies to all consumption curves rising with increasing speed. The increased fuel consumption when driving at different speeds also increases due to additional consumption due to acceleration. This yields the following energy-saving rule: in order to minimize energy consumption, the forklift should move at a constant speed that is needed to maintain the required time to travel the route.

If it moves along individual sections L_i with the use of fuel in the tank purchased at different prices p_{Bi}, the cost of fuel at different speeds in each section vt results from the following Eq. (4.72):

$$K_B(V_i) = \sum p_{Bi} \cdot C_B(V_i) \cdot L_i = \sum p_{Bi}\left(C_0 + C_1 \cdot V_i^n\right) \cdot L_i \tag{4.72}$$

Unlike fuel consumption, the cost of that fuel at different speeds in each section can be optionally lower than in case of constant speed. This means an additional opportunity to lower costs, which can be used to optimize profit. If the fuel price is the same for the entire route, the fuel costs are proportional to the fuel consumption (4.70); therefore, they are the lowest when driving at a constant speed. Then the Eq. (4.72) for the dependence of speed on the cost of fuel at a fixed price of fuel pB and the total path length L suggests as follows (Fig. 4.28):

Transport time and maximum cargo load

$$T_{Tnm} = \sum\left(T_{Hi} + \frac{L_i}{Vi}\right) = N_H \cdot t_H + \frac{L}{V_m} \tag{4.73}$$

L means the total path V_m the average speed of movement (4.73), while t_H means the average waiting time according to the formula:

$$t_H = \left(\frac{1}{N_H}\right) \cdot \sum t_{Hi} \tag{4.74}$$

The standby time is the total time allocated for the slowdown, entrance, deposition/retrieval of the freight unit, acceleration and possible waiting.

For the regular movement of a freight unit by a forklift between two points in a warehouse with no stops along the way, where the total road L = 2 − s according to Eq. (4.74), the transport time will be Tr = 2 − (tH + s/v). Operating data suggest a time dependence of the transport time and the speed of movement, as shown in Fig. 4.29. In the case of speed reduction from 25 km/h by 20 % to 20 km/h, the

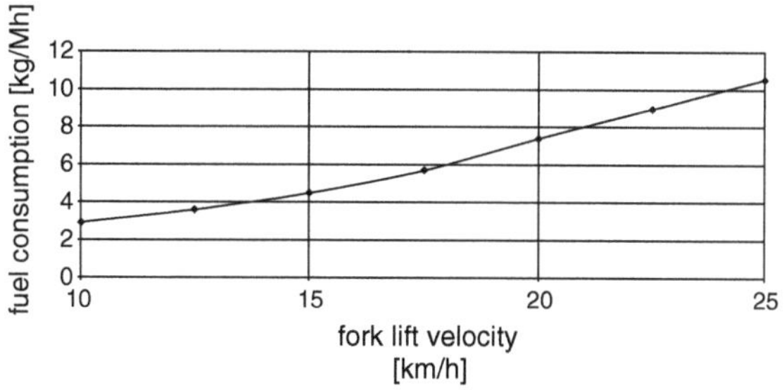

Fig. 4.28 The dependence of fuel consumption on speed for an exemplary forklift truck

transport time is extended by 25 %. Reducing speed by half to 12.5 km/h will lead to almost double the transport time.

The total duration of the travel T_u[h] for a combined cycle of the forklift's operation to N_h storage areas and with section lengths Li between individual storage areas H_i and H_{i+1} is also apparent from Eq. (4.74), wherein the sum refers to all storage areas, to which the forklift commutes i = 1, 2, … Nh. Therefrom with the frequency of the forklift's travel f_u over a longer operation period T_e, e.g. one year = 360 d = 8640, follows the following equation:

$$f_U(V_m) = \frac{T_E}{N_H.t_H + \frac{L}{V_m}}$$

With the effective forklift capacity C_{eff} this results with the limit load $F_s(v_m)$ in the following equation:

$$F_S(V_m) = C_{eff} \cdot f_U = \frac{C_{eff} \cdot T_E}{\left(N_H \cdot t_H + \frac{L}{V_m}\right)}$$

The effective capacity of the truck (e.g. a tractor with several trailers) is its factory capacity C multiplied by the maximum filling indicator pmax, which means that Ceff = pmax-C. In the case of routes to and fro, without stops (return to stop-standby) for a one-way path s, NS = 2, L = 2 − s and the load limit FS (vm) = Ceff-TE/2(tH + s/vm). From the graph in Fig. 4.8 it can be read that reducing the speed from 25 km/h by 20 % down to 20 km/h allows lowering the load limit load by 18 %, while reducing speed by half also almost halves the cargo load limit to 22 %. Therefore, the load limit increases and decreases proportionally to speed, provided the total driving time is significantly higher than the total time at standby. The number of forklifts N_S needed to carry a certain amount of goods F_a between specified storage areas, wherein the said amount of goods is significantly higher than the truck load limit, that value is N_s = ROUNDED(F_a/F_s).

Therefore, the number of forklift trucks that is needed for a large amount of cargo, is growing, jumping in integers, almost in proportion to speed. Reducing speed with large amounts of cargo to be transported requires the use of more

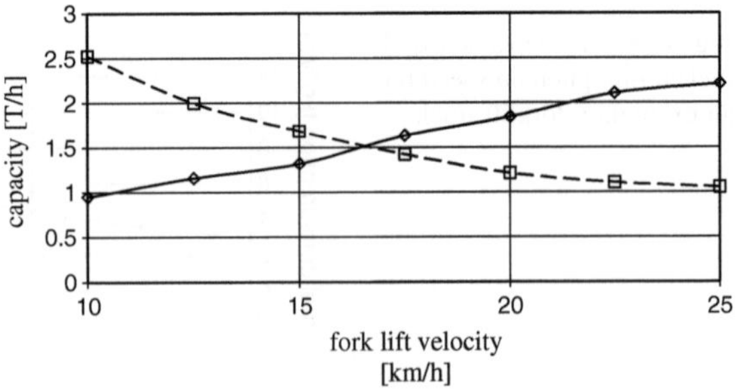

Fig. 4.29 The relation of time and speed and carrying capacity and speed [311]

forklifts. If the amount of cargo to be transported drops at constant number of forklifts and speed N_S, at their load limit N_S–F_S, then the average capacity utilization pS = FA/(N_s–F_s) will falls below 100 %. Then a reduction of speed will reduce the limit load, causing better capacity utilization and lower operating costs.

Chapter 5
Experimental Research Results

Apart from [82], no test results on the energy consumption of moving a palette in actual facilities have been published. They are also published in statistical yearbooks, for example [283]. Sorry to say—this is understandable, since warehouses, LSM, usually do not keep records of energy consumption, particularly in terms of a single freight unit going through the LSM. From the point of view of organization, such information is fairly easy to gather in a working warehouse equipped only with forklifts, whose working time is recorded e.g. for the Office of Technical Inspection, or for maintaining hardware warranty. Adequate record keeping could give information on the energy intensity of moving a single freight unit. For other devices, the problem is similar. Unfortunately, companies do not have such statements.

Analyzing the state of the energy intensity of the LSM, it was assumed that it will be carried out in a large world class company, which has eight warehouses (hub-type) in Poland, adapted for handling pallet unit with dimensions 800 × 1200 × 18,000 mm. Storage facilities are located in logistics parks of a single operator, and therefore these facilities are similar to one another other in terms of organizational and technical structure. It was therefore assumed that there will be no need for additional statistical processing of the data collected.

In all facilities, the company uses Komatsu and Jungheinrich forklift trucks [291], and has been operating them from purchase for about 10 years. Monitoring fuel consumption was conducted through the recording of the energy intensity of a specific forklift and of each freight unit individually.

All transport tasks carried out in the warehouse are recorded in a WMS computer system in conjunction with the carrier (pallet)—which is already the standard for WMS systems, and at the same time a prerequisite for the job [304, 305].

When it comes to fuel economy standards during operation, specified by the manufacturer (depending on the load status of the truck) in the range of 3.2–4.0 l/h, although the forklift in fact consumed 4.5 l/h, while another type of forklift: 4.8-5.8 l/h, while in reality it consumed 3.87 l/h.

In the course of the study, it has been observed that the energy intensity of the truck operated by different operators in the same tasks differed by up to 30 %. Also, in addition to various fuel consumption, the same type of forklift, working under the

P. Zajac, *Evaluation Method of Energy Consumption in Logistic Warehouse Systems*, EcoProduction, DOI 10.1007/978-3-319-22044-4_5

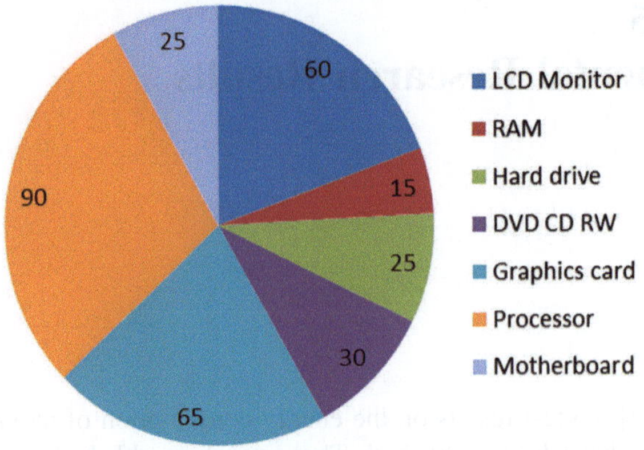

Fig. 5.1 Summary of the energy intensity of individual components of an office computer

same conditions, operated by different operators required less or more frequent workshop repairs, e.g. replacement of rear tires (sometimes the front ones as well).

When using the ABC method for the frequency of retrievals [138] to allocation of storage areas, freight units showed an average energy intensity of moving the pallet through the warehouse: 80,750.00 J, with differences between freight units of up to 60 %.

However, this value should be increased by the energy intensity of information interchange systems—offices and Auto-ID systems, EDI, WMS and office lighting. In the case of warehouses located in logistics parks, the standard module is 8000 m^2. There is usually an office there, with approx. 60 m^2, with office equipment (Fig. 5.1) and computer hardware, including automatic identification, EDI, WMS, lighting and heating during winter (in accordance with the common existing standards), which requires the supply of approximately 8415 kWh per year.

The mean total energy intensity of moving a freight unit through the warehouse is: 80,950.00 J.

Using the model for assessing energy intensity, this value was reduced by 9 % in terms of the energy intensity of handling equipment.

Chapter 6
Verification of the Model Evaluation of LSM

This section presents the application of this model on three examples (items 6.1., 6.2., 6.3., respectively):

- The energy intensity of the dynamic storage technology in the LSM was rated,
- The energy consumption of LSM was rated, with varied techniques of information interchange: a desktop scanner was used alternatively with a docking station and radio communication,
- LSM was rated in terms of energy intensity, alternating between forklifts with properly selected operating parameters in LSM, with different driving characteristics of the powertrain.

6.1 Evaluation of Dynamic Energy Intensity LSM

Evaluation of the energy intensity of logistic storage systems in dynamic storage technology uses:

- Tunnel racks with replaceable fork carrier DIS-2 Jungheinrich,
- Tunnel racks with satellite BT Radioshuttle forklift,
- Drive-in racks,
- Flow racks,
- Drive-through racks.

The energy intensity of these storage technologies largely depends on the adopted queuing system. If drive-in racks are supported only on one side, then there is a LIFO or FILO queue—and where they are loaded on one side and at the same time unloaded on the other, then the FIFO method is applied.

Drive-in and drive-through racks are designed for storing large and heavy goods of narrow range. Both types of racks have the advantages of static and block storage, resulting in maximum utilization of space thanks to the considerable height of the rack and the security (e.g. against crushing) of the goods stored.

Prior to entering the rack—aisle, the forklift raises the pallet to a certain storage height. The forklift itself cannot be wider than the pallet, therefore this purpose is

P. Zajac, *Evaluation Method of Energy Consumption in Logistic Warehouse Systems*, EcoProduction, DOI 10.1007/978-3-319-22044-4_6

best suited by forklifts with sideways placement of the operator with respect to the direction of travel.

A tunnel rack with DIS-2 (Drive-In-System) or (BT-System), pallets are transported in tunnels by means of self-propelled forklift carriages (carts in case of BT), which move into the channel (level) of storage. They are driven by individual drive assemblies and equipped with batteries as a power source and are computer-controlled: they use the WMS and/or operator panel from the forklift.

These storage techniques are used because corridors takes less space available in the warehouse than in other cases. Moreover, it is possible to get a relatively large filling of shelves with a very diverse range of goods.

The application area of tunnel rack systems with interchangeable DIS-2 fork carrier or satellite carriage is similar to drive-in and drive-through racks.

The DIS-2 can be operated by any forklift with an FEM carriage, and a single truck can operate several mobile carriages—while the forklift and the replaceable fork carriage are connected together permanently.

A forklift with an integrated self-propelled carriage takes a palette. It deposits it to the allocated rack corridor. Then using the operator terminal, the self-propelled carriage mechanism is launched, which drives the pallet to the end of rack or the last pallet already placed in the rack. The carriage then goes back and is received by the forklift and can perform further actions. In the meantime, the forklift can handle other carriages in other corridors.

The mechanism for collecting pallets from the corridor is similar. An empty carriage is parked by forklift at the end of the corridor and is started using the terminal. The carriage reaches the last pallet placed in the corridor, picks it up and goes back to the beginning of the corridor. The self-propelled carriage with the pallet is received by the forklift and transported to its destination.

Designers of logistics storage systems can choose from among the options available on the market: solutions with a tunnel/satellite cart e.g. Linde, Still, ISL—Pallet Mole, Nedcon—Pallet Shuttle, but the most popular and the oldest is BT Radioshuttle.

The BT Radioshuttle system does not feature self-propelled carriages—they were replaced with transport carts, on which a pallet is placed in the racking corridor.

It is not possible to simultaneously transport the BT Radioshuttle cart, so placing the pallet in the racking corridor requires two passes with a forklift.

As a test facility for the calculations a warehouse of one of the Polish retail chains was selected. The freight unit has a size of 800 × 1200 × 1800 mm. The weight together with the pallet is 750 kg. Storage takes place on 5 levels (0 + 4). The total storage height (the last beam of the rack) is 8400 mm. The racks are designed to hold 10870 pallet pieces; however, according to the specific nature of DIS-2, it uses 9783 pallet slots. In this case, using a reach truck to handle drive-in racks, in order to reach the same boundary conditions, 8 pieces of carts should be acquired and the surface should be expanded, because the storage height in racks allows only using about 70 % of the area; also in this case, the warehouse will hold 13976 pallet pieces (Fig. 6.1).

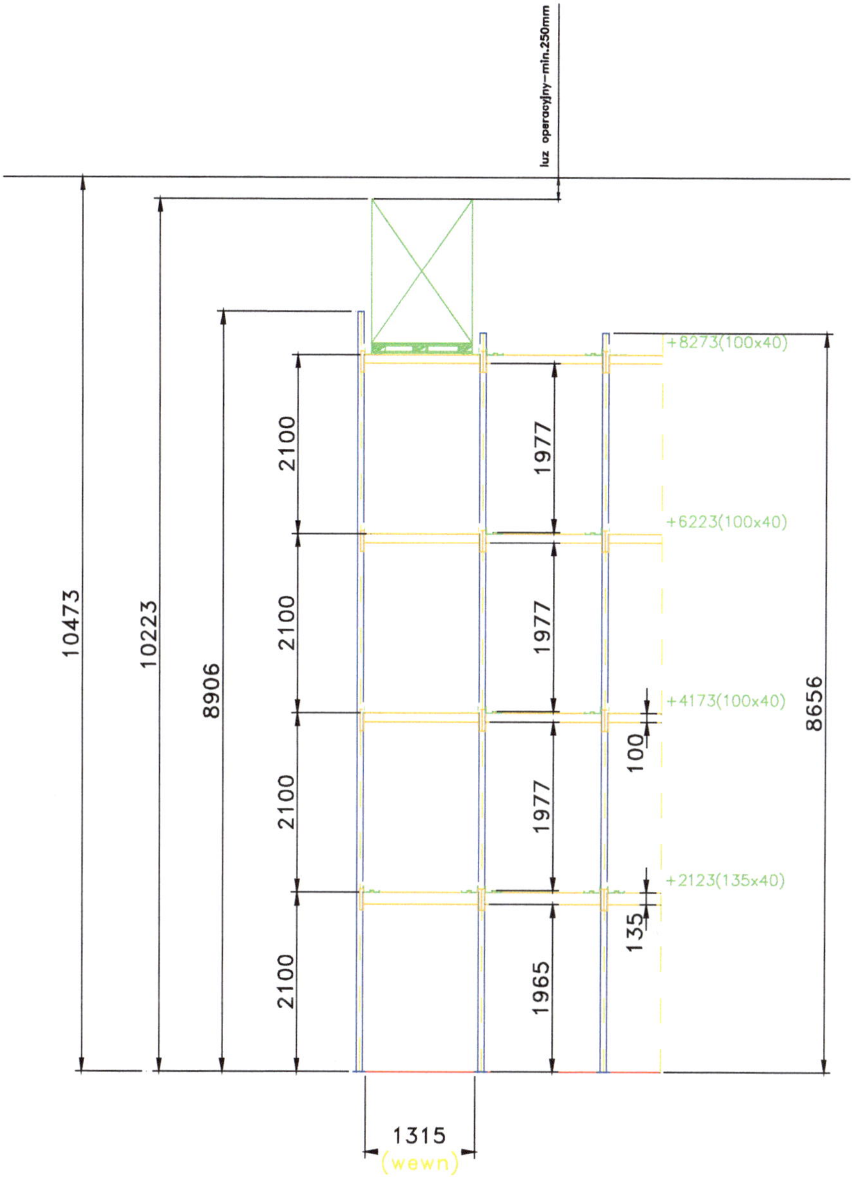

Fig. 6.1 Diagram for a shelf rack with characteristic dimensions for: DIS-2, RadioSchutlle, drive-in. Author's own work

It was assumed in the comparative calculations that forklifts are ETV 214 (their parameters are further used in the calculation). Racks adapted to BT Radioshuttle are more expensive than racks able to handle DIS-2, which in practice is of great importance when deciding on the choice of the system.

Table 6.1 List of parameters

	DIS-2	BT Radio shuttle	Drive-in racks	Flow racks	Sliding racks
MP capacity	2520	2520	2100	2040	2160
Filling level %	90	90	70	98	98
MP utilization	2268	2268	1470	1999	2117
€/MP	56	64	46	127	117
€ Vehicles	56,240	56,240	33,240	33,240	33,240
Investment/M.P. utilization	87	96	88	146	135
Warehouse costs 1 m^2 = € 610	247	247	381	280	265
Invest. cost/MP	334	343	469	426	400
Energy intensity of use pallets (MJ)	7873.25	9715.00	8175.68	5768.35	86007.18

Author's own work

Table 6.1 shows a comparison of the parameters of the above-mentioned shelving systems and the forklifts supporting them. Apart from BT, other systems used shelving and carts made by Jungheinrich. Tabela_radioshuttle_DIS2.

Analyzing the results (Table 6.1), we can conclude that, with the same surface, the DIS-2 and BT Radioshuttle systems can accommodate the most pallets. Other systems: drive-in racks, flow racks and sliding racks—can accommodate approx. 400 pallets fewer. The designers further analyze the filling ratio of shelves. In this context, the most preferred are flow racks and sliding racks. The commonly used drive-in shelf shows rather weak performance.

An important factor is the cost of the investment converted to a single storage slot—and here drive-ins turn out to be most profitable, with the DIS-2 and BT Radioshuttle systems being only a little more expensive. Flow racks and sliding racks are incomparably more expensive.

For the sake of comparison, it is assumed that the costs of forklift trucks used in the DIS-2 and BT Radioshuttle system are the same. The same level of costs of forklift trucks is also assumed for classical solutions: drive-in, flow and sliding racks. The cost of purchasing forklifts working with drive-in, flow and sliding racks is lower than for use in DIS-2 and BT Radioshuttle systems—DIS-2 and BT Radioshuttle requires equipping forklifts with carriage cranes (DIS-2) or carts, on which the pallet is placed in the racking corridor (BT Radioshuttle). As practice shows, in the case of tunnel rack technique and removable fork carriage, the "critical mass" of freight units is 500 pcs.—and more pallet slots in the warehouse.

6.2 Evaluation of Energy Consumption of the IT System

This section presents the impact of the IT subsystem on the energy intensity of moving a freight unit through LSM. The case is unique in the literature, because it tells the story of only 8 warehouses owned by a single company, which underwent revitalization in the subsequent stages of their operation. Usually in such situations, no one pays attention to the energy consumption of information systems, or their impact on the energy intensity of LSM. Nevertheless, we were successful in gaining results for LSM energy intensity when using (Figs. 6.2, 6.3 and Table 6.2):

- barcode readers 1D (warehouse 1),
- barcode readers integrated with portable terminals—solution with a docking station (warehouse 2),
- barcode readers integrated with portable terminals—solution with a radio terminal (warehouse 3).

The results of the calculations were compared with results from actual facilities. Data for comparative calculations:

- Handling documents related to one pallet takes 2 min 30 s. This time includes: performance of activities related to the data entry into the warehouse computer system, printing orders, forwarding the order to the forklift operator (this assumption applies to warehouse 1—supported by barcodes).

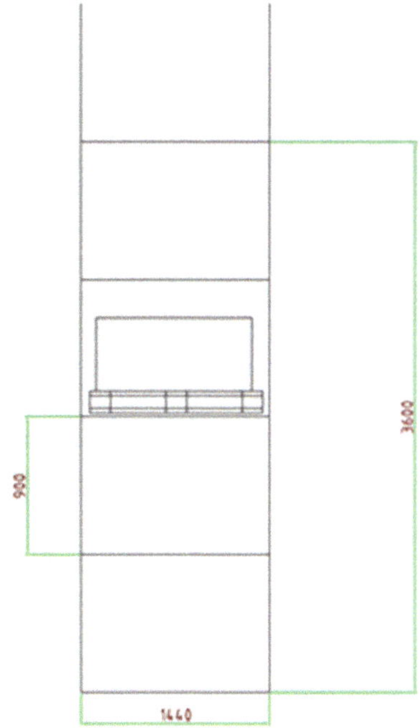

Fig. 6.2 Construction sketch of a shelving slot. Author's own work

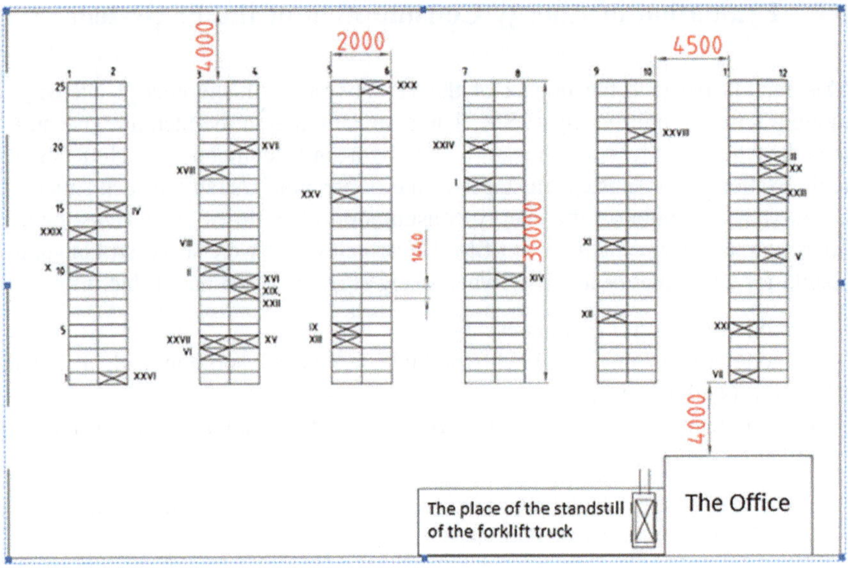

Fig. 6.3 Layout of the warehouse with dimensions. Author's own work

Table 6.2 Warehouse design parameters

Shelving slots	Number of racks	Number of levels	Number of rows	Number of slots per row
1500	6	5	12	25

- In the case of warehouses 2 and 3, the activities related to the transfer of information on the transport of pallets take 60 s. This time is used by the operator to read the information from the system.
- The G ratio (total time consumption of transport contacts during the year) was calculated as the product of the number of pallets in the warehouse and the time required for the filling out documents related to in-house transport. It was then increased by 35 % (author's assumption).
- All other coefficients are the percentage share of G (depending on the type of warehouse).
- The warehouses have the same annual turnover (according to the above amounting to 400 thousand pallets), they differ by their information interchange system.

The warehouse is made up of six double-sided racks (12 rows), consisting of five levels, on which pallets can be stoed. In each row there are 25 shelving slots from the front of the rack.

- All forklifts operating in the territory of warehouses are the same
- Forklifts and any related characteristics have been derived from the Jungheinrich model EFG 316 (characteristics and dimensions shown in Fig. 6.4)
- There are five forklifts operating in the warehouse, but for clarity of calculations, the author assumes that all the work they perform follow each other sequentially (as if only one forklift was working)
- A forklift moves at a constant speed of 13 km/h in a straight line
- A forklift slows down to 5 km/h while cornering
- Shelving slots are marked by giving their row number, the number of the slot in a row and the number of the level at which the cargo pallet is placed
- In the case of warehouse 2, docking terminals are located on the edges and in the middle of rows
- For simplicity, the author assumes that all pallets handled are of the same weight of 800 kg (weight of cargo with pallet)
- The study consists in simulated transport of 30 pallets from random locations to gates (randomly selected by the author)
- The simulation is carried out for the same routes for the three models of warehouses
- The mechanical energy required to perform the tasks is given by the following formula:

The forklift starts its route from the place designated as a standby area, then head for the pallet with the cargo (random place), retrieves it, drives it to one of the three receiving gates (at random) after which it takes a spot in the standby area, where it receives another order. These steps are repeated 30 times.

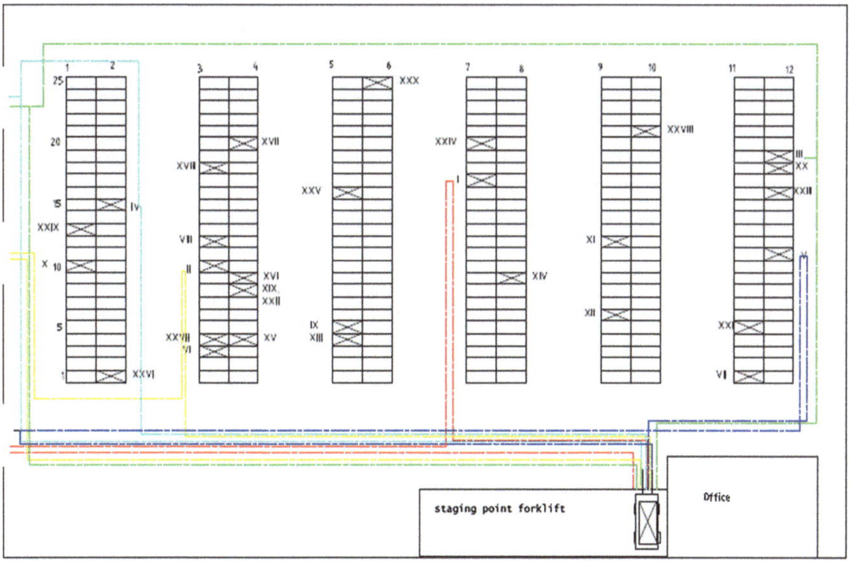

Fig. 6.4 Diagram of forklift's travel in warehouse 1 (routes 1–5)

Table 6.3 Comparative results

Warehouse 1 (m)	Warehouse 2 (m)	Warehouse 3 (m)
3637.1	2595.5	2339.8

Above is a diagram of a forklift route for warehouse 1. To illustrate the transport process the travel routes for the first five transport orders were shown.

To illustrate the process of transport, the routes for the first five transport orders are shown.

Warehouse 2

Forklift operating characteristics: in this case, the forklift's route is changed. This is due to the fact that there are docking slots mounted in shelves, which allows the operator to read their current transport orders. The Author assumes that orders are available every time the operator connects their device. Docking stations are located at the edges of shelves and half their length.

The route to the docking terminal is marked in green. Other colors represent routing for orders.

Warehouse 3

In this case, the warehouse is equipped with a device for transferring electronic messages. In practice, this means that the operator receives a transport order directly to the terminal installed in the forklift. In this type of solution, there is no need to move to the docking device (warehouse 2), which allows a further minimization of the travel route.

Summary of the three distances the forklift has travelled depending on the information interchange system use (Table 6.3).

Earlier assumptions imply that the forklift moves along straight lines at a constant speed of 13 km/h. During cornering, however, this speed decreases to 5 km/h. For greater clarity, the calculations assume that the path the forklift needs to go round corners, slow down to 5 km/h and re-accelerate to 13 km/h is 6 m (counted in a straight line).

The total energy used by a forklift during transport tasks (Tables 6.4 and 6.5).

The introduction of an electronic information interchange system does not increase the energy intensity of LSM by more than 8 % of the total energy consumption of LSM; in fact, it reduces the LSM energy consumption—by up to 27 % in the analyzed cases.

Table 6.4 Summary of total energy calculations for individual warehouses

Energy intensity of LSM			
	Warehouse 1	Warehouse 2	Warehouse 3
$E_k + Ep$ (MJ)	17.26	13.45	12.75

Table 6.5 List of indicators of significance of introducing the system of electronic documents interchange

Energy intensity of EDI			
	Warehouse 1	Warehouse 2	Warehouse 3
Ew **edi** (MJ)	0.1364	0.0884	0.0524

6.3 The Use of Model Evaluation of Energy Consumption in LSM Management

Analyzing the method developed in [137, 138, 144, 145], one will notice that reaching the optimal solution for LSM takes place through successive solutions, while a longer—several years' worth of analysis on a single LSM will lead to the optimal solution. However, this will be the case only when it is assumed that the strategy of allocation remains the same. This also applies to the characteristics of machines and handling equipment. Moreover, finding the optimal solution according to known methods is true for the approved and ongoing picking program, the type and quantity of products and the volume of stocks.

Assumptions regarding the available resources		
Number of forklift operators	18	Pc
Number of available forklifts	6	Pc
Number of shifts per day	3	
Number of hours in one shift	7.5	H
Number of working days per year	302	
The turnover on cargo pallet storage units per year	500.000	Pc
The turnover on cargo pallet storage units per day	1656	Pc

In this case, for the new operating conditions LSM, which strongly change the pre-design state, one can use the developed methods for assessing the energy intensity of LSM. Below is an example illustrating such a case.

As shown in Table 6.6, with an annual turnover of 500.000 pieces and at 18 forklift operators working in three shifts, the time buffer per one pallet load is 15 min.

Table 6.7 presents the speeds to choose from for the DOOSAN B20X-5 forklift, whose characteristics were presented in Fig. 1.3, while driving along specific sections of the route (Table 6.8).

Table 6.6 Summary of data and figures relating to the warehouse and the spacing of shelves

	Marking	Value	Unit
Distance between the wall and the edge of the shelf	v	2	m
Distance between transport tracks and the shelf	Y	1	m
Distance between transport tracks	a	4	m
Number of columns	8		pcs
Number of rows	2		pcs

Author's own work

The above summary of data and figures relating to the warehouse and the spacing of shelves indications are shown in Fig. 6.7.

Table 6.7 Summary data relating to values associated with the forklift truck and the cargo

	Marking	Value	Unit
Load weight	ml	1000	kg
Truck weight	m2	300	kg
Truck weight with load	m3	1300	kg
Cargo width	b	1.3	m
Cargo length	c	1.8	m
Cargo height	d	1.5	m
Maximum speed of the forklift	v_{max}	13	km/h
Cornering speed	v_{min}	5	km/h

Author's own work

The warehouse layout is shown in Figs. 6.7 and 6.8, which is a common one

Table 6.8 Summary of speeds achieved by the forklift along different sections of the route

Forklift truck DOOSAN B20X-5	Symbol	Value	Unit
Maximum speed achieved while driving the main corridors	V_{max1}	13	km/h
The maximum permitted speed when driving picker corridors	V_{max2}	8	km/h
Speed reached on bends	V_{min}	5	km/h

Author's own work

Case 1

The first case shows the energy consumption when driving the truck at maximum speeds. Tables 6.3 and 6.4 show the power consumption at appropriate speeds, as well as the corresponding sections of the route, and the graphs in Figs. 6.5 and 6.6 show these values in graphical form (Fig. 6.7, Tables 6.9 and 6.10).

With the increase in speed, the energy consumption also increases. While moving the truck with a load when the truck goes to the storage area, the first corner is on the 0.008 km stretch, which is why the highest speed that the truck can reach with the specified parameters is 7 km/h, and will be reached at 0.003 km. The truck travels at a constant speed of over 3 m, and then slows down to the adopted cornering speed, which is 5 km/h. This speed is achieved at a distance of 8 meters. It then accelerates again, this time reaching a maximum speed of 8 km/h. This result is possible because the path before the next turn is longer and is 45 m in this case. This time, however, it slows down to the speed of 0 km/h, where it stops completely at the selected storage area.

On its way back, the truck reaches maximum speeds, is lighter as it moves without additional weight, it reaches these speeds in less time, and this causes it to move at a constant speed longer. As can be seen in Fig. 6.6, the energy consumed while driving the truck without load is smaller than movement with the load, despite the fact that this truck is moving longer at maximum speed. This is mainly because the weight of the vehicle is lower. Tables 6.11 and 6.12 show the most important values pertaining to this case.

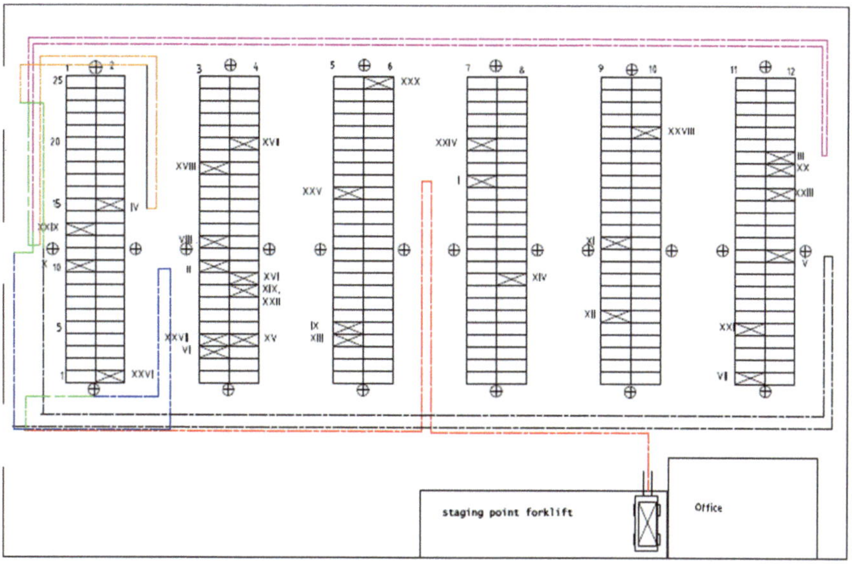

Fig. 6.5 A view of the forklift's travel in warehouse 2 for the first five courses

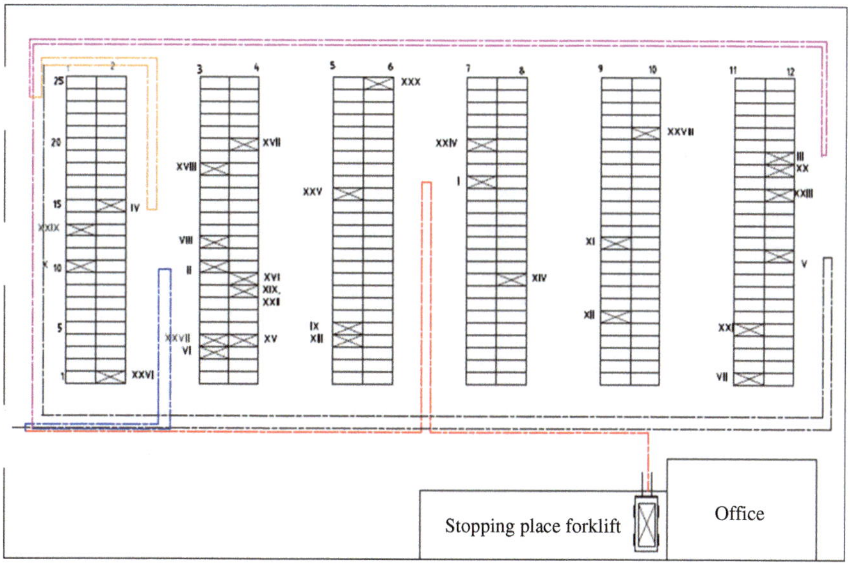

Fig. 6.6 Illustration of a forklift travel (for the first five orders) in Warehouse

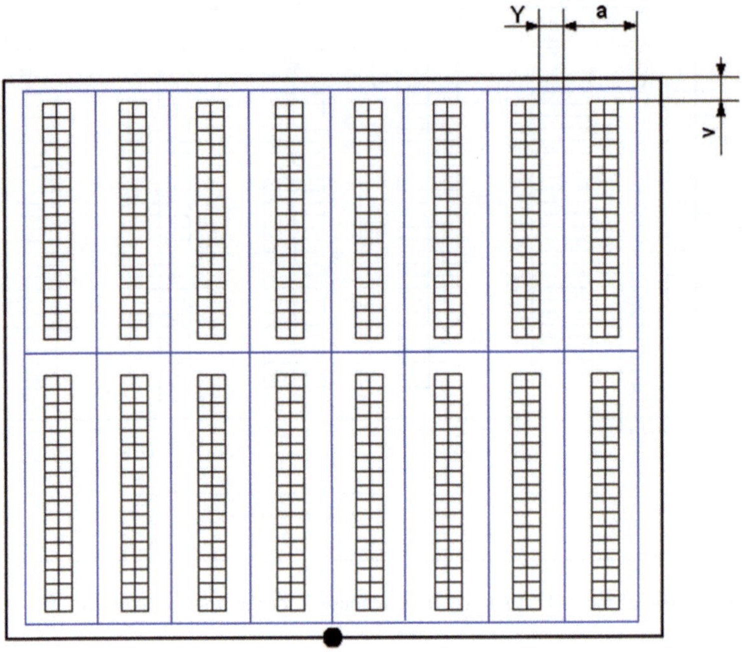

Fig. 6.7 Summary of data and figures relating to the warehouse and the spacing of shelves. Author's own work

Table 6.9 The dependence of energy consumption on the speed and the routes travelled for a truck with load (V_{max})

MAS (km)	V (km/h)	MWt2 (MJ)
0.00001	0	0
0.00006	1	0.01
0.00015	2	0.04
0.00030	3	0.08
0.00055	4	0.12
0.00095	5	0.16
0.00167	6	0.2
0.00326	7	0.22
0.00688	7	0.23
0.00760	6	0.23
0.00800	5	0.23
0.00872	6	0.26
0.01031	7	0.28
0.01509	8	0.29
0.04934	8	0.29
0.05092	7	0.29
0.05164	6	0.29
0.05205	5	0.29
0.05229	4	0.29
0.05245	3	0.29
0.05254	2	0.29
0.05258	1	0.29
0.05260	0	0.29

Author's own work

Table 6.10 The dependence of energy consumption on the speed and the routes travelled for truck without load (V_{max})

MAS (km)	V (km/h)	MWt1 (MJ)
0.00000	0	0
0.00001	1	0.01
0.00004	2	0.04
0.00007	3	0.08
0.00013	4	0.12
0.00022	5	0.16
0.00038	6	0.2
0.00074	7	0.22
0.04438	8	0.22
0.04474	7	0.22
0.04491	6	0.22
0.04500	5	0.22
0.04517	6	0.25
0.04553	7	0.27
0.05262	7	0.27
0.05278	6	0.27
0.05287	5	0.27
0.05293	4	0.27
0.05296	3	0.27
0.05299	2	0.27
0.05300	1	0.27
0.05300	0	0.27

Author's own work

Case 2

In this variant, the energy intensity of LSM differs from "Case 2" in that the speed of the forklift while cornering is the same as in straight sections. Tables 6.9 and 6.10 present the distributions of energy consumption for this case.

The energy consumption at the appropriate speeds, as well as the corresponding sections of the route presented in Tables 6.13 and 6.14. In this case, the vehicle takes 1 s to reach the speed, at which it will move all the way, and will travel the entire route in nearly 40 s, or nearly twice as long as in the first example.

Tables 6.15 and 6.16 show the dependence of the routes traversed in time, assuming that the stroller moves with constant speed the whole time—from the moment of reaching the speed of 5 km/h up to a complete halt.

The results indicate that the work time increases twice when driving at minimum speed, as compared to moving the same truck at maximum speed. This is a feasible solution, because the result does not exceed the permissible time spent per pallet unit; meanwhile, it turns out that the energy consumption in this case is also a considerably smaller. Table 6.17 summarizes the results relating to the cost, time, energy consumed and the costs associated with driving the truck at different speeds.

Table 6.11 Energy consumption of a forklift while driving at maximum speeds with load

V (km/h)	V (m/s)	Vśr (m/s)	FN(V) or FH(V) (N)	rb (N/kN)	p(V) (N/kN)	AS (km)	MAS (km)	At (s)	MAt (s)	MAt (min)	AW (MJ)	MWt (MJ)
0	0.00	0.14	4500	46.80	299.35	0.00001	0.00001	0.00	0.00	0.00000	0.000	0.000
1	0.28	0.42	4000	46.85	260.85	0.00005	0.00006	0.17	0.17	0.00005	0.014	0.014
2	0.56	0.69	3600	46.94	229.98	0.00009	0.00015	0.16	0.34	0.00009	0.029	0.043
3	0.83	0.97	3100	47.09	191.37	0.00015	0.00030	0.18	0.52	0.00014	0.039	0.081
4	1.11	1.25	2600	47.29	152.71	0.00025	0.00055	0.22	0.74	0.00021	0.043	0.124
5	1.39	1.53	2100	47.55	113.99	0.00040	0.00095	0.29	1.03	0.00029	0.041	0.164
6	1.67	1.81	1600	47.86	75.21	0.00072	0.00167	0.43	1.46	0.00041	0.033	0.197
7	1.94	1.94	1140	48.23	39.46	0.00158	0.00326	0.81	2.28	0.00063	0.020	0.217
7	1.94	1.81	1140	48.23	39.46	0.00362	0.00688	1.86	4.14	0.00115	0.009	0.226
6	1.67	1.53	1600	47.86	75.21	0.00072	0.00760	0.43	4.57	0.00127	0.000	0.226
5	1.39	1.53	2100	47.55	113.99	0.00040	0.00800	0.29	4.86	0.00135	0.000	0.226
6	1.67	1.81	1600	47.86	75.21	0.00072	0.00872	0.43	5.29	0.00147	0.033	0.259
7	1.94	2.08	1140	48.23	39.46	0.00158	0.01031	0.81	6.11	0.00170	0.020	0.279
8	2.22	2.22	825	48.65	14.81	0.00478	0.01509	2.15	8.26	0.00229	0.009	0.288
8	2.22	2.08	825	48.65	14.81	0.03425	0.04934	15.41	23.67	0.00658	0.001	0.289
7	1.94	1.81	1140	48.23	39.46	0.00158	0.05092	0.81	24.49	0.00680	0.000	0.289
6	1.67	1.53	1600	47.86	75.21	0.00072	0.05164	0.43	24.92	0.00692	0.000	0.289
5	1.39	1.25	2100	47.55	113.99	0.00040	0.05205	0.29	25.21	0.00700	0.000	0.289
4	1.11	0.97	2600	47.29	152.71	0.00025	0.05229	0.22	25.43	0.00706	0.000	0.289
3	0.83	0.69	3100	47.09	191.37	0.00015	0.05245	0.18	25.61	0.00711	0.000	0.289
2	0.56	0.42	3600	46.94	229.98	0.00009	0.05254	0.16	25.78	0.00716	0.000	0.289
1	0.28	0.14	4000	46.85	260.85	0.00005	0.05258	0.17	25.95	0.00721	0.000	0.289
0	0.00	0.00	4500	46.80	299.35	0.00001	0.05260	0.00	25.95	0.00721	0.000	0.289

Author's own work

Table 6.12 Energy consumption of a forklift while driving at maximum speeds without load

V (km/h)	V (m/s)	Vśr (m/s)	FN(V) or FH(V) (N)	AS (km)	MAS (km)	At (s)	MAt (s)	MAt (min)	AW (MJ)	MWt (MJ)
0	0.00	0.14	4500	0.00000	0.00000	0.00000	0.00000	0.00000	0.00	0.00
1	0.28	0.42	4000	0.00001	0.00001	0.03975	0.03975	0.00001	0.01	0.01
2	0.56	0.69	3600	0.00002	0.00004	0.03756	0.07731	0.00002	0.03	0.04
3	0.83	0.97	3100	0.00004	0.00007	0.04211	0.11942	0.00003	0.04	0.08
4	1.11	1.25	2600	0.00006	0.00013	0.05085	0.17027	0.00005	0.04	0.12
5	1.39	1.53	2100	0.00009	0.00022	0.06652	0.23679	0.00007	0.04	0.16
6	1.67	1.81	1600	0.00017	0.00038	0.09902	0.33580	0.00009	0.03	0.20
7	1.94	2.08	1140	0.00036	0.00074	0.18532	0.52112	0.00014	0.02	0.22
8	2.22	2.08	825	0.04364	0.04438	19.63700	20.15812	0.00560	0.00	0.22
7	1.94	1.81	1140	0.00036	0.04474	0.18532	20.34345	0.00565	0.00	0.22
6	1.67	1.53	1600	0.00017	0.04491	0.09902	20.44246	0.00568	0.00	0.22
5	1.39	1.53	2100	0.00009	0.04500	0.06652	20.50898	0.00570	0.00	0.22
6	1.67	1.81	1600	0.00017	0.04517	0.09902	20.60799	0.00572	0.03	0.25
7	1.94	1.94	1140	0.00036	0.04553	0.18532	20.79332	0.00578	0.02	0.27
7	1.94	1.81	1140	0.00709	0.05262	3.64655	24.43986	0.00679	0.00	0.27
6	1.67	1.53	1600	0.00017	0.05278	0.09902	24.53888	0.00682	0.00	0.27
5	1.39	1.25	2100	0.00009	0.05287	0.06652	24.60540	0.00683	0.00	0.27
4	1.11	0.97	2600	0.00006	0.05293	0.05085	24.65624	0.00685	0.00	0.27
3	0.83	0.69	3100	0.00004	0.05296	0.04211	24.69835	0.00686	0.00	0.27
2	0.56	0.42	3600	0.00002	0.05299	0.03756	24.73591	0.00687	0.00	0.27
1	0.28	0.14	4000	0.00001	0.05300	0.03975	24.77567	0.00688	0.00	0.27
0	0.00	0.00	4500	0.00000	0.05300	0.00000	24.77567	0.00688	0.00	0.27

Author's own work

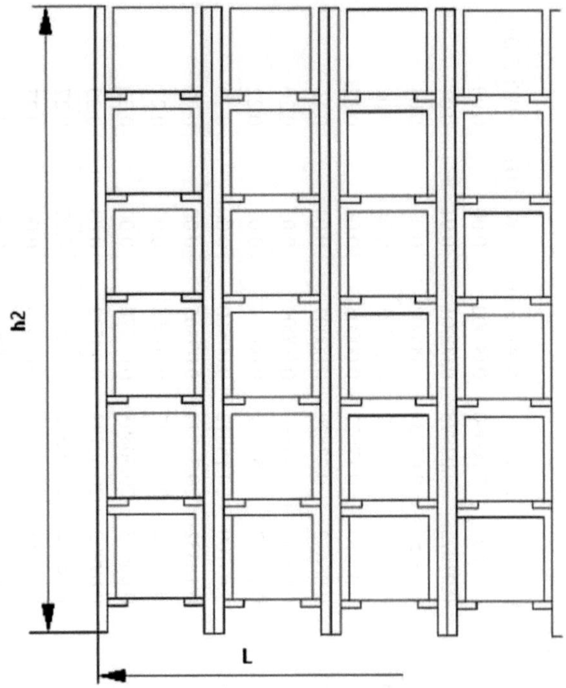

Fig. 6.8 Schematic of a rack section. Author's own work

Table 6.13 The dependence of energy consumption on the speed and the route travelled by truck with load (V_{min})

$\sum \Delta S$ (km)	V (km/h)	$\sum Wt2$ (MJ)
0.00001	0	0.00000
0.00006	1	0.01370
0.00015	2	0.04258
0.00030	3	0.08116
0.00055	4	0.12372
0.00095	5	0.16434
0.05245	5	0.16475
0.05270	4	0.16475
0.05285	3	0.16475
0.05294	2	0.16475
0.05299	1	0.16475
0.05300	0	0.16475

Author's own work

Table 6.17 shows that the best solution for this case is to move the vehicle at minimum speed, that is 5 km/h. The time the forklift takes to travel this route is less than the buffer time per pallet unit, which is why this solution carries the least cost, while not disturbing the organization and operation of the facility.

Table 6.14 The dependence of energy consumption on the speed and the route travelled by truck without load (V_{min})

$\sum \Delta S$ (km)	V (km/h)	$\sum Wt2$ (MJ)
0.00000	0.00000	0.00000
0.00001	1.00000	0.01428
0.00004	2.00000	0.04377
0.00007	3.00000	0.08301
0.00013	4.00000	0.12627
0.00022	5.00000	0.16764
0.05203	5.00000	0.16836
0.05208	4.00000	0.16836
0.05212	3.00000	0.16836
0.05214	2.00000	0.16836
0.05215	1.00000	0.16836
0.05215	0.00000	0.16836

Author's own work

Table 6.18 presents the results of calculations of energy intensity of LSM for speeds V_{min} and V_1.

These considerations suggest that reducing the travel time to the speed Vmin would reduce the energy intensity of LSM but for that to be acceptable, the criterion of the time window and the number of forklift operators must be met.

The presented case will be profitable with the forklift operator's remuneration of up to 4.80PLN/h (formula below):

$$\frac{\text{cost difference}}{\text{difference\_between\_truck\_operators} * \text{working\_time}} = \frac{108731}{10 * 302\,\text{days} \;\; * 7.5\,\text{h}} = 4.80 \tag{6.1}$$

Case 3

Tables 6.19 and 6.20 show the distribution of energy consumption during truck operation at maximum speed, while Tables 6.21 and 6.22 present work while driving the truck at minimum speeds.

These parameters allow the vehicle to reach higher speeds in a shorter time, which leads to travelling the route faster, but the energy is several times higher than in the previous case. Pooled results are presented in Table 6.12.

Table 6.23. Summary results for the Tables 6.24 and 6.25 show the distribution of energy consumption during truck operation at maximum speed, while Tables 6.26 and 6.27 show the operation while driving at minimum speeds.

Table 6.15 Energy consumption when driving a forklift truck at minimum speed with load

V (km/h)	V (m/s)	Vśr (m/s)	FN(V) or FH(V) (N)	rb (N/kN)	p(V) (N/kN)	ΔS (km)	∑ΔS (km)	Δt (s)	∑ Δt(s)	∑ Δt (min)	ΔW (MJ)	∑ Wt(MJ)
0	0.00	0.14	4500	46.80	299.35	0.00001	0.00001	0.00000	0.00000	0.00000	0.00	0.00
1	0.28	0.42	4000	46.85	260.85	0.00005	0.00006	0.17260	0.17260	0.00005	0.01	0.01
2	0.56	0.69	3600	46.94	229.98	0.00009	0.00015	0.16314	0.33574	0.00009	0.03	0.04
3	0.83	0.97	3100	47.09	191.37	0.00015	0.00030	0.18298	0.51872	0.00014	0.04	0.08
4	1.11	1.25	2600	47.29	152.71	0.00025	0.00055	0.22112	0.73984	0.00021	0.04	0.12
5	1.39	1.39	2100	47.55	113.99	0.00040	0.00095	0.28965	1.02949	0.00029	0.04	0.16
5	1.39	1.25	2100	47.55	113.99	0.05150	0.05245	37.07737	38.10686	0.01059	0.00	0.16
4	1.11	0.97	2600	47.29	152.71	0.00025	0.05270	0.22112	38.32799	0.01065	0.00	0.16
3	0.83	0.69	3100	47.09	191.37	0.00015	0.05285	0.18298	38.51097	0.01070	0.00	0.16
2	0.56	0.42	3600	46.94	229.98	0.00009	0.05294	0.16314	38.67410	0.01074	0.00	0.16
1	0.28	0.14	4000	46.85	260.85	0.00005	0.05299	0.17260	38.84670	0.01079	0.00	0.16
0	0.00	0.00	4500	46.80	299.35	0.00001	0.05300	0.00000	38.84670	0.01079	0.00	0.16

Author's own work

Table 6.16 Energy consumption when driving a forklift truck at minimum speed without load

V (km/h)	V (m/s)	Vśr (m/s)	FN(V) or FH (V) (N)	rb (N/kN)	p(V) (N/kN)	ΔS (km)	∑ΔS (km)	Δt (s)	∑Δt (s)	∑Δt (min)	ΔW (MJ)	∑Wt (MJ)
0	0.00	0.14	4500	200.65	1299.35	0.00000	0.00000	0.00000	0.00000	0.00000	0.00	0.00
1	0.28	0.42	4000	200.78	1132.55	0.00001	0.00001	0.03975	0.03975	0.00001	0.01	0.01
2	0.56	0.69	3600	201.15	998.85	0.00002	0.00004	0.03756	0.07731	0.00002	0.03	0.04
3	0.83	0.97	3100	201.75	831.59	0.00004	0.00007	0.04211	0.11942	0.00003	0.04	0.08
4	1.11	1.25	2600	202.58	664.09	0.00006	0.00013	0.05085	0.17027	0.00005	0.04	0.13
5	1.39	1.39	2100	203.64	496.36	0.00009	0.00022	0.06652	0.23679	0.00007	0.04	0.17
5	1.39	1.25	2100	203.64	496.36	0.05181	0.05203	37.30051	37.53729	0.01043	0.00	0.17
4	1.11	0.97	2600	202.58	664.09	0.00006	0.05208	0.05085	37.58814	0.01044	0.00	0.17
3	0.83	0.69	3100	201.75	831.59	0.00004	0.05212	0.04211	37.63025	0.01045	0.00	0.17
2	0.56	0.42	3600	201.15	998.85	0.00002	0.05214	0.03756	37.66781	0.01046	0.00	0.17
1	0.28	0.14	4000	200.78	1132.55	0.00001	0.05215	0.03975	37.70756	0.01047	0.00	0.17
0	0.00	0.00	4500	200.65	1299.35	0.00000	0.05215	0.00000	37.70756	0.01047	0.00	0.17

Author's own work

Table 6.17 Summary of results

	$\sum$ Wt (MJ) with load	$\sum$ Wt (MJ) without load	$\sum$ Wt (kWh) with load	$\sum$ Wt (kWh) without load	Cost of travel[a] (PLN)	$\sum \Delta t$ (s) with load	$\sum \Delta t$ (s) without load
V_{max}	0.28929	0.27331	0.08100	0.07653	0.02518	25.94945	24.77567
V_{min}	0.16475	0.16836	0.04613	0.04714	0.01551	38.84670	37.70756
Difference	0.12453	0.10495	0.03487	0.02939	0.00967	−12.89725	−12.93190

[a]It was assumed that the cost of electricity consumption is 0.3219PLN/1kWh [20]

Table 6.18 Summary of the difference in costs and energy used

	Value	Value	Unit	Difference
	V_{min}	V_1		
Average route to the storage area	545.4117647	545.4117647	m	
Number of forklift trucks needed	32	22	pcs	−10
Energy used	1751190.30	2931515.18	MJ	1180324.9
Cost of maintaining the warehouse	161319.65	270051.18	PLN	108731.53

Author's own work

Table 6.19 Energy consumption of forklift (2) while running at maximum speeds with load

V (km/h)	V (m/s)	FN(V) or FH(V) (N)	ΔS (km)	∑Δs (km)	Δt (s)	∑Δt (s)	ΔW (MJ)	∑Wt (MJ)
0	0.00	9000	0.00000	0.00000	0.00	0.00	0.00	0.00
1	0.28	8100	0.00001	0.00002	0.05	0.05	0.05	0.05
2	0.56	7200	0.00003	0.00004	0.05	0.09	0.10	0.15
3	0.83	6300	0.00004	0.00008	0.05	0.14	0.14	0.30
4	1.11	5400	0.00006	0.00015	0.06	0.20	0.17	0.46
5	1.39	4500	0.00009	0.00024	0.07	0.27	0.17	0.64
6	1.67	3600	0.00015	0.00039	0.09	0.35	0.16	0.80
7	1.94	2900	0.00022	0.00060	0.11	0.47	0.15	0.94
8	2.22	2200	0.00036	0.00096	0.16	0.63	0.12	1.06
9	2.50	1600	0.00066	0.00162	0.26	0.89	0.08	1.14
10	2.78	1200	0.00126	0.00288	0.45	1.34	0.05	1.19
11	3.06	900	0.00239	0.00527	0.78	2.13	0.03	1.23
10	2.78	1200	0.00126	0.00652	0.45	2.58	0.00	1.23
9	2.50	1600	0.00066	0.00718	0.26	2.84	0.00	1.23
8	2.22	1600	0.00036	0.00754	0.16	3.00	0.00	1.23
7	1.94	2900	0.00022	0.00776	0.11	3.12	0.00	1.23
6	1.67	3600	0.00015	0.00791	0.09	3.20	0.00	1.23
5	1.39	4500	0.00009	0.00800	0.07	3.27	0.00	1.23
6	1.67	3600	0.00015	0.00815	0.09	3.36	0.16	1.39
7	1.94	2900	0.00022	0.00836	0.11	3.47	0.15	1.53
8	2.22	2200	0.00036	0.00872	0.16	3.63	0.12	1.65
8	2.22	2200	0.04358	0.05230	19.61	23.24	0.00	1.65
7	1.94	2900	0.00022	0.05252	0.11	23.35	0.00	1.65
6	1.67	3600	0.00015	0.05267	0.09	23.44	0.00	1.65
5	1.39	4500	0.00009	0.05276	0.07	23.51	0.00	1.65
4	1.11	5400	0.00006	0.05282	0.06	23.57	0.00	1.65
3	0.83	6300	0.00004	0.05286	0.05	23.61	0.00	1.65
2	0.56	7200	0.00003	0.05289	0.05	23.66	0.00	1.65
1	0.28	8100	0.00001	0.05290	0.05	23.71	0.00	1.65
0	0.00	9000	0.00000	0.05291	0.00	23.71	0.00	1.65

Author's own work

Table 6.20 Energy consumption of forklift (2) while running at maximum speeds without load

V (km/h)	V (m/s)	FN(V) or FH(V) (N)	ΔS (km)	∑Δs (km)	Δt (s)	∑Δt (s)	ΔW (MJ)	∑Wt (MJ)
0	0.00	9000	0.00000	0.00000	0.00	0.00	0.00	0.00
1	0.28	8100	0.00000	0.00000	0.01	0.01	0.05	0.05
2	0.56	7200	0.00001	0.00001	0.01	0.02	0.10	0.15
3	0.83	6300	0.00001	0.00002	0.01	0.03	0.14	0.30
4	1.11	5400	0.00001	0.00003	0.01	0.05	0.17	0.46
5	1.39	4500	0.00002	0.00006	0.02	0.06	0.17	0.64
6	1.67	3600	0.00003	0.00009	0.02	0.08	0.16	0.80
7	1.94	2900	0.00005	0.00014	0.03	0.11	0.15	0.95
8	2.22	2200	0.00008	0.00022	0.04	0.14	0.12	1.06
8	2.22	2200	0.00767	0.00789	3.45	3.60	0.00	1.06
7	1.94	2900	0.00005	0.00794	0.03	3.62	0.00	1.06
6	1.67	3600	0.00003	0.00798	0.02	3.64	0.00	1.06
5	1.39	4500	0.00002	0.00800	0.02	3.66	0.00	1.06
6	1.67	3600	0.00015	0.00815	0.09	3.75	0.04	1.10
7	1.94	2900	0.00022	0.00836	0.11	3.86	0.03	1.13
8	2.22	2200	0.00036	0.00872	0.16	4.02	0.03	1.16
9	2.50	1600	0.00066	0.00938	0.26	4.28	0.02	1.18
10	2.78	1200	0.00126	0.01064	0.45	4.74	0.01	1.19
11	3.06	900	0.00309	0.01373	1.01	5.75	0.01	1.20
12	3.33	700	0.02141	0.03514	6.42	12.17	0.00	1.20
12	3.33	700	0.01205	0.04719	3.61	15.79	0.00	1.20
11	3.06	900	0.00309	0.05028	1.01	16.80	0.00	1.20
10	2.78	1200	0.00126	0.05154	0.45	17.25	0.00	1.20
9	2.50	1600	0.00066	0.05220	0.26	17.51	0.00	1.20
8	2.22	2200	0.00036	0.05256	0.16	17.68	0.00	1.20
7	1.94	2900	0.00022	0.05278	0.11	17.79	0.00	1.20
6	1.67	3600	0.00015	0.05292	0.09	17.88	0.00	1.20
5	1.39	4500	0.00002	0.05294	0.02	17.89	0.00	1.20
4	1.11	5400	0.00001	0.05296	0.01	17.90	0.00	1.20
3	0.83	6300	0.00001	0.05297	0.01	17.92	0.00	1.20
2	0.56	7200	0.00001	0.05297	0.01	17.93	0.00	1.20
1	0.28	8100	0.00000	0.05298	0.01	17.94	0.00	1.20
0	0.00	9000	0.00000	0.05298	0.00	17.94	0.00	1.20

Author's own work

Case 4

The last Case presents the work done for forklift (3), whose drive force values for individual speeds are lower than in the previous cases (Table 6.28).

Table 6.21 Energy consumption of forklift (2) while driving at minimum speed with load

V (km/h)	V (m/s)	FN(V) or FH(V) (N)	ΔS (km)	∑Δs (km)	Δt (s)	∑Δt (s)	ΔW (MJ)	∑Wt (MJ)
0	0.00	9000	0.00000	0.00000	0.00	0.00	0.00	0.00
1	0.28	8100	0.00001	0.00002	0.05	0.05	0.05	0.05
2	0.56	7200	0.00003	0.00004	0.05	0.09	0.10	0.15
3	0.83	6300	0.00004	0.00008	0.05	0.14	0.14	0.30
4	1.11	5400	0.00006	0.00015	0.06	0.20	0.17	0.46
5	1.39	4500	0.00009	0.00024	0.07	0.27	0.17	0.64
5	1.39	4500	0.05261	0.05285	37.88	38.15	0.00	0.64
4	1.11	5400	0.00006	0.05292	0.06	38.20	0.00	0.64
3	0.83	6300	0.00004	0.05296	0.05	38.25	0.00	0.64
2	0.56	7200	0.00003	0.05298	0.05	38.30	0.00	0.64
1	0.28	8100	0.00001	0.05300	0.05	38.35	0.00	0.64
0	0.00	9000	0.00000	0.05300	0.00	38.35	0.00	0.64

Author's own work

Table 6.22 Energy consumption of forklift (2) while driving at minimum speed without load

V (km/h)	V (m/s)	FN(V) or FH(V) (N)	ΔS (km)	∑Δs (km)	Δt (s)	∑Δt (s)	ΔW (MJ)	∑Wt (MJ]
0	0.00	9000	0.00000	0.00000	0.00	0.00	0.00	0.00
1	0.28	8100	0.00000	0.00000	0.01	0.01	0.05	0.05
2	0.56	7200	0.00001	0.00001	0.01	0.02	0.10	0.15
3	0.83	6300	0.00001	0.00002	0.01	0.03	0.14	0.30
4	1.11	5400	0.00001	0.00003	0.01	0.05	0.17	0.46
5	1.39	4500	0.00002	0.00006	0.02	0.06	0.17	0.64
5	1.39	4500	0.05269	0.05274	37.93	38.00	0.00	0.64
4	1.11	5400	0.00001	0.05276	0.01	38.01	0.00	0.64
3	0.83	6300	0.00001	0.05277	0.01	38.02	0.00	0.64
2	0.56	7200	0.00001	0.05277	0.01	38.03	0.00	0.64
1	0.28	8100	0.00000	0.05277	0.01	38.04	0.00	0.64
0	0.00	9000	0.00000	0.05278	0.00	38.04	0.00	0.64

Author's own work

6.4 Method to Assess the Energy Consumption of LSM with the Use of RESOLVER

Using the model for evaluation energy intensity is associated with a fairly large amount of calculations; therefore for the purpose of the model, an advisory system was developed based on the standard platform of the RESOLVER advisory system.

The advisory system allows the user to set a task in LSM, taking into account the criterion of energy intensity, dialoguing with the operator (sample data attached

Table 6.23 Energy consumption of forklift (3) while running at maximum speeds with load

	$\sum$Wt (MJ) with load	$\sum$Wt (MJ) without load	$\sum\Delta$t (min) with load	$\sum\Delta$t (min) without load
V_{max}	1.65018	1.20136	23.70828	17.93741
V_{min}	0.63681	0.63747	38.34718	38.04149
Difference	1.01337	0.56389	−14.63890	−20.10408

Author's own work.hird example. Author's own work

Table 6.24 Energy consumption of forklift (3) while running at maximum speeds with load

V (km/h)	V (m/s)	FN(V) or FH(V) (N)	ΔS (km)	$\sum\Delta$s (km)	Δt (s)	$\sum\Delta$t (s)	ΔW (MJ)	$\sum$Wt (MJ)
0	0.00	4000	0.00001	0.00001	0.00	0.00	0.00	0.00
1	0.28	3300	0.00004	0.00005	0.13	0.13	0.02	0.02
2	0.56	2700	0.00008	0.00013	0.14	0.28	0.03	0.05
3	0.83	2200	0.00015	0.00027	0.18	0.45	0.04	0.09
4	1.11	1800	0.00025	0.00053	0.23	0.68	0.04	0.13
5	1.39	1400	0.00047	0.00099	0.34	1.02	0.03	0.17
6	1.67	1100	0.00090	0.00190	0.54	1.56	0.03	0.19
7	1.94	800	0.00288	0.00478	1.48	3.04	0.01	0.20
7	1.94	800	0.00232	0.00710	1.19	4.23	0.01	0.22
6	1.67	1100	0.00090	0.00800	0.54	4.77	0.00	0.22
5	1.39	1400	0.00047	0.00847	0.34	5.11	0.00	0.22
6	1.67	1100	0.00090	0.00937	0.54	5.65	0.03	0.24
7	1.94	800	0.00288	0.01225	1.48	7.14	0.01	0.26
7	1.94	800	0.03932	0.05157	20.22	27.36	0.00	0.26
6	1.67	1100	0.00090	0.05247	0.54	27.90	0.00	0.26
5	1.39	1400	0.00047	0.05294	0.34	28.24	0.00	0.26
4	1.11	1800	0.00025	0.05319	0.23	28.46	0.00	0.26
3	0.83	2200	0.00015	0.05334	0.18	28.64	0.00	0.26
2	0.56	2700	0.00008	0.05342	0.14	28.78	0.00	0.26
1	0.28	3300	0.00004	0.05346	0.13	28.91	0.00	0.26
0	0.00	4000	0.00001	0.05347	0.00	28.91	0.00	0.26

Author's own work.hird example. Author's own work

below). The advisory system operates according to the decision-making diagram set out in Figs. 6.9 and 6.10.

Sample screenshots of the advisory computer system are shown in Figs. 6.11a, b.

For the purposes of test calculations, specifications were prepared and introduced into the system regarding the following forklifts: **Still, Linde and Jungheinrich.** The advisory system applied the *Fuzzy logic, allowing for the construction of expert systems* close to actual solutions. At the beginning, all trucks were recorded in the database in accordance with the formula below:

Table 6.25 Energy consumption of forklift (3) while running at maximum speed without load

V (km/h)	V (m/s)	FN(V) or FH(V) (N)	ΔS (km)	∑Δs (km)	Δt (s)	∑Δt (s)	ΔW (MJ)	∑Wt (MJ)
0	0.00	4000	0.00000	0.00	0.00	0.00	0.00	0.00
1	0.28	3300	0.00001	0.00	0.03	0.03	0.02	0.02
2	0.56	2700	0.00002	0.00	0.03	0.06	0.03	0.05
3	0.83	2200	0.00003	0.00	0.04	0.10	0.04	0.09
4	1.11	1800	0.00006	0.00	0.05	0.16	0.04	0.13
5	1.39	1400	0.00011	0.00	0.08	0.23	0.04	0.17
6	1.67	1100	0.00021	0.00	0.12	0.36	0.03	0.19
7	1.94	800	0.00064	0.00	0.33	0.68	0.01	0.21
8	2.22	600	0.04309	0.04	19.39	20.07	0.00	0.21
7	1.94	800	0.00064	0.04	0.33	20.40	0.00	0.21
6	1.67	1100	0.00021	0.05	0.12	20.52	0.00	0.21
5	1.39	1400	0.00011	0.05	0.08	20.60	0.00	0.21
6	1.67	1100	0.00021	0.05	0.12	20.72	0.03	0.23
7	1.94	800	0.00064	0.05	0.33	21.05	0.01	0.24
7	1.94	800	0.00672	0.05	3.46	24.51	0.00	0.25
6	1.67	1100	0.00021	0.05	0.12	24.63	0.00	0.25
5	1.39	1400	0.00011	0.05	0.08	24.71	0.00	0.25
4	1.11	1800	0.00006	0.05	0.05	24.76	0.00	0.25
3	0.83	2200	0.00003	0.05	0.04	24.80	0.00	0.25
2	0.56	2700	0.00002	0.05	0.03	24.84	0.00	0.25
1	0.28	3300	0.00001	0.05	0.03	24.87	0.00	0.25
0	0.00	4000	0.00000	0.05	0.00	24.87	0.00	0.25

Author's own work

1.	Manufacturer brand
2.	Type
3.	Type of drive
4.	Maximum lift height
5.	Maximum capacity
6.	Corridor width
7.	Gradeability
8.	Maximum speed with load
9.	Weight
10.	Maximum thrust with load

System operation is illustrated below: in order to accurately perform calculations of energy consumption, each truck has been described with an increment of speed by 0.1 km/h from the start up to the maximum value specified by the manufacturer.

Table 6.26 Energy consumption of forklift (3) while driving at minimum speed of cargo

V (km/h)	V (m/s)	FN(V) or FH(V) (N)	ΔS (km)	$\sum$ Δs (km)	Δt (s)	$\sum$ Δt (s)	ΔW (MJ)	$\sum$ Wt (MJ)
0	0.00	4000	0.00001	0.00001	0.00	0.00	0.00	0.00
1	0.28	3300	0.00004	0.00005	0.13	0.13	0.02	0.02
2	0.56	2700	0.00008	0.00013	0.14	0.28	0.03	0.05
3	0.83	2200	0.00015	0.00027	0.18	0.45	0.04	0.09
4	1.11	1800	0.00025	0.00053	0.23	0.68	0.04	0.13
5	1.39	1400	0.00047	0.00099	0.34	1.02	0.03	0.17
5	1.39	1400	0.05148	0.05247	37.07	38.08	0.00	0.17
4	1.11	1800	0.00025	0.05273	0.23	38.31	0.00	0.17
3	0.83	2200	0.00015	0.05287	0.18	38.49	0.00	0.17
2	0.56	2700	0.00008	0.05295	0.14	38.63	0.00	0.17
1	0.28	3300	0.00004	0.05299	0.13	38.76	0.00	0.17
0	0.00	4000	0.00001	0.05300	0.00	38.76	0.00	0.17

Author's own work

Table 6.27 Energy consumption of forklift (3) while driving at minimum speed without load

V (km/h)	V (m/s)	FN(V) or FH(V) (N)	ΔS (km)	$\sum$ Δs (km)	Δt (s)	$\sum$ Δt (s)	ΔW (MJ)	$\sum$ Wt (MJ)
0	0.00	4000	0.00000	0.00000	0.00000	0.00000	0.00000	0.00000
1	0.28	3300	0.00001	0.00001	0.03070	0.03070	0.01774	0.01774
2	0.56	2700	0.00002	0.00003	0.03292	0.06362	0.03301	0.05075
3	0.83	2200	0.00003	0.00006	0.04039	0.10402	0.04033	0.09108
4	1.11	1800	0.00006	0.00012	0.05210	0.15612	0.04169	0.13277
5	1.39	1400	0.00011	0.00023	0.07697	0.23309	0.03529	0.16805
5	1.39	1400	0.05184	0.05207	37.32671	37.55980	0.00017	0.16822
4	1.11	1800	0.00006	0.05213	0.05210	37.61190	0.00000	0.16822
3	0.83	2200	0.00003	0.05216	0.04039	37.65230	0.00000	0.16822
2	0.56	2700	0.00002	0.05218	0.03292	37.68522	0.00000	0.16822
1	0.28	3300	0.00001	0.05219	0.03070	37.71592	0.00000	0.16822
0	0.00	4000	0.00000	0.05219	0.00000	37.71592	0.00000	0.16822

Author's own work

After calculating all the parameters for each truck, they have been summarized in a single graph showing the increase in energy consumption along with the path traversed (Tables 6.29 and 6.30).

The final selection is made based on the plot summary of all the models that meet the given criteria.

The above summary shows the difference in energy consumption between electric and engine-powered forklifts. The difference in this case is fairly constant (very slight fluctuations occur with increasing path travelled).

Table 6.28 Summary of results for forklift (3)

	$\sum$ Wt (MJ) with load	$\sum$ Wt (MJ) without load	$\sum \Delta$t (min) with load	$\sum \Delta$t (min) without load
V_{max}	0.25645	0.24584	28.91478	24.86680
V_{min}	0.16757	0.16822	38.76252	37.71592
Difference	0.08888	0.07762	−9.84774	−12.84912

Author's own work

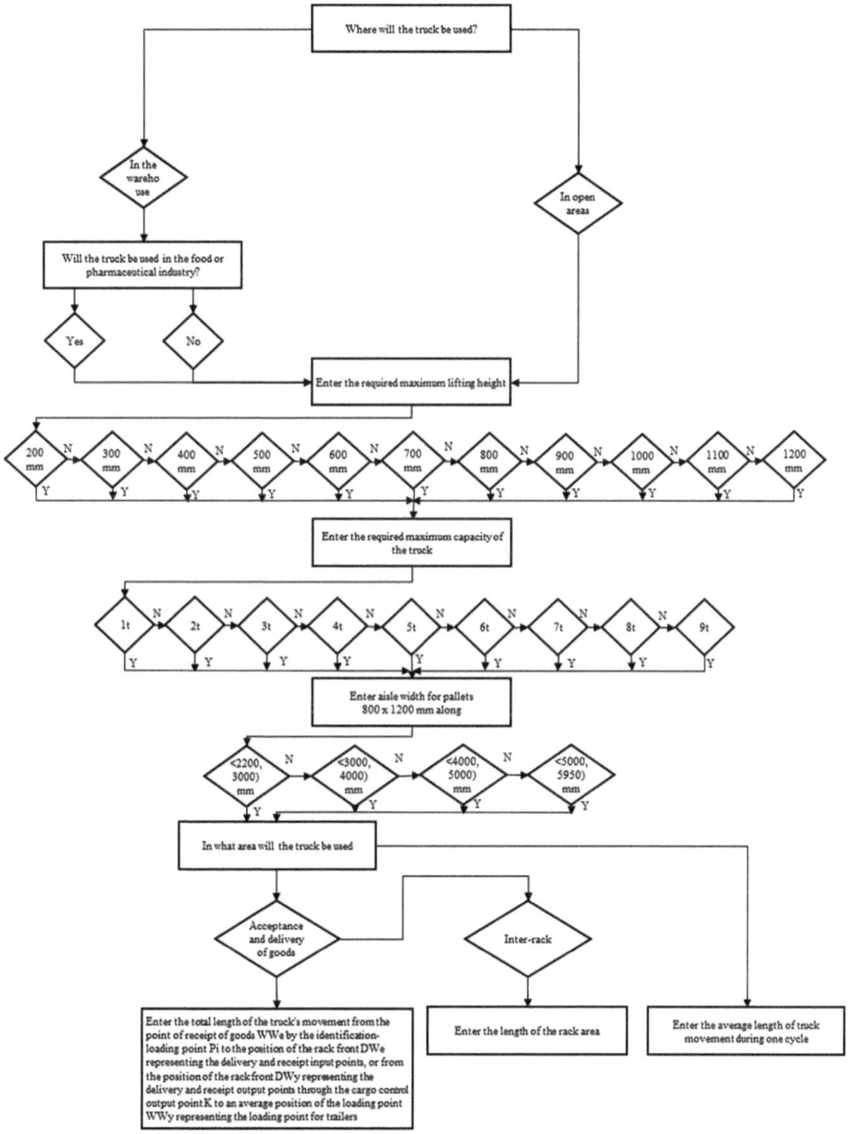

Fig. 6.9 Diagram of the decision-making process of the advisory system part 1

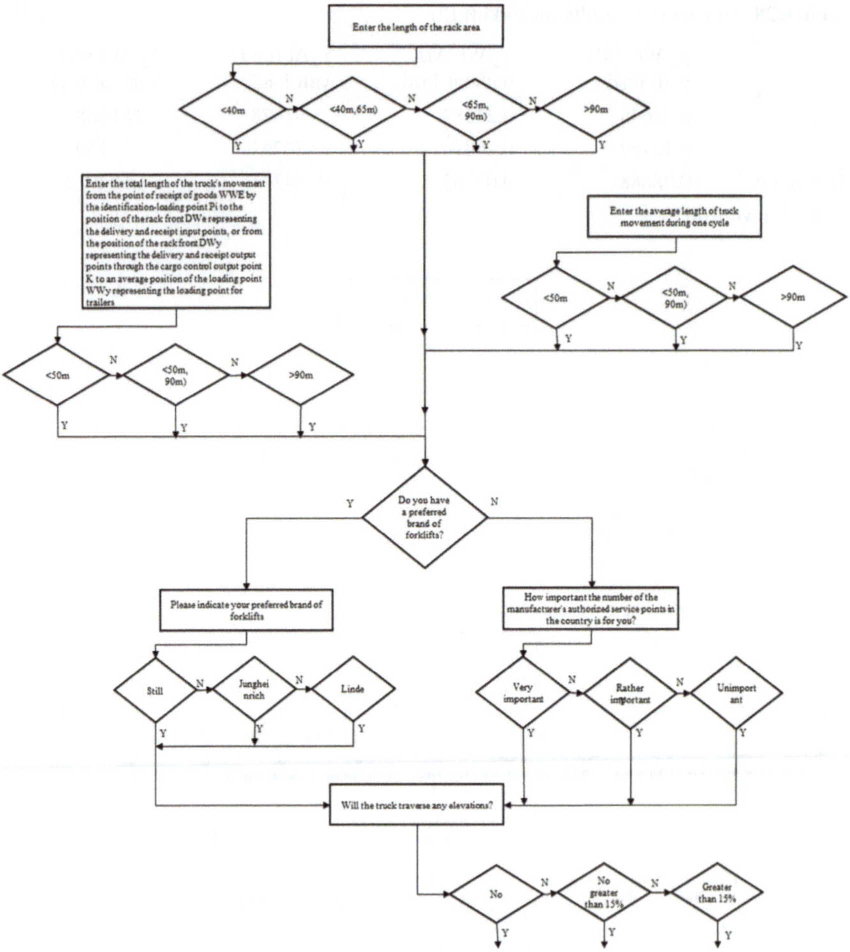

Fig. 6.10 Diagram of the decision-making process of the advisory system part 2

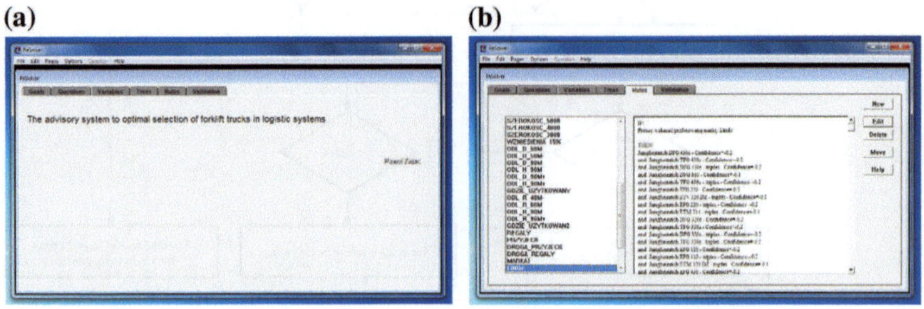

Fig. 6.11 **a** Home page of the advisory system in ReSolver and **b** Case of rules in the expert system

Table 6.29 Sample summary of the distance traveled by the truck, and the corresponding amount of energy used for an engine-powered truck—Jungheinrich TFG 320s, operated in open space

m	44.6	55.8	90.7	149.9
MJ	0.0593	0.0667	0.0894	0.1279

Table 6.30 The percentage ratio of energy consumed by an electric truck—Linde E30 and by an engine-powered truck—Jungheinrich TFG 430s

Jungheinrich TFG 430s	0.0898	0.0987	0.1319	0.1877	W (MJ)
Linde E30	0.0561	0.0613	0.0821	0.1192	
Ratio	160.1	161.0	160.7	157.5	%

In summary, calculation results of the advisory system show that:

- When it comes to energy intensity, electric forklifts are by far the most prefered. Compared to the gas-powered forklifts (still the most popular in Poland), they take up to 60 % less energy while maintaining the same capacity. With the added value of energy recovery during braking—this makes up to 10 % of the total energy used.
- Among the currently used engine-powered forklifts, trucks powered by LPG are preferred over diesel. In most cases, they consume less energy than their diesel counterparts, although a huge difference can be seen only in the cost of ownership of both vehicles.

Summary and Conclusions

Based on the results obtained in the paper, the following conclusions are formulated:

1. The model of energy intensity of LSM presented in the paper allows specifying a "map" of energy intensity of any logistic storage system.
2. The model of LSM energy intensity allows calculating the energy intensity of moving a freight unit through the LSM, comparing the energy intensity of moving a freight unit in the area of a warehouse or warehouses.
3. Analyzing the indirect results of using the model by means of successive iterations, one can achieve optimal device configuration for manual loading of freight units as well as their operating parameters—primarily the speed of movement of the freight units.
4. This method of evaluating energy intensity using the model allows evaluating the energy consumption of individual processes as well as the logistics storage system.

The allocation of work areas in LSM for handling equipment can also be adjusted in view of the drive characteristics of devices, and thus their suitability for use under appropriate conditions.

Employees and managers in LSM have been observed to have poor awareness of the issue of energy intensity, and thus only the faintest interest in collecting and analyzing data in terms of reducing the energy intensity of LSM. Promoting the model in enterprises can reduce the energy intensity of the national economy. This proposal was presented to the President of the Republic of Poland Bronislaw Komorowski as part of the project, which was positively assessed by Professors: Professor Assoc. W. Rydzkowski (University of Gdansk), Prof. SGH Assoc. H. Brdulak (School of Economics, Warsaw).

In the future, the model will need expansion based on the results of various types of storage, including refrigerated warehouses, freezers, and those in which heat interchange takes place (e.g. fresh fruit storage).

P. Zajac, *Evaluation Method of Energy Consumption in Logistic Warehouse Systems*, EcoProduction, DOI 10.1007/978-3-319-22044-4

Utilitarian Results

The utilitarian result of this paper is seen in the proposed "map" of the energy intensity of LSM, which, used by LSM managers or designers, as well as professionals interested in controlling the energy intensity of LSM (e.g. in the context of cost), would allow for the optimization of energy intensity, thus reducing the energy intensity of the global economy. Moreover, it complements the already known method of space allocation for SKUs or can be an alternative to methods that use scheduling in warehouse processes.

The system developed in RESOLVER for the evaluation of energy intensity, allowing for a fairly quick assessment of any LSM means that LSM operation can be carried out in terms of energy intensity.

P. Zajac, *Evaluation Method of Energy Consumption in Logistic Warehouse Systems*, EcoProduction, DOI 10.1007/978-3-319-22044-4

Appendix 1
Database of the Expert System

P. Zajac, *Evaluation Method of Energy Consumption in Logistic Warehouse Systems*, EcoProduction, DOI 10.1007/978-3-319-22044-4

Lp.	Manufacturer	Type	Drive	Lifting height (mm)	Capacity (t)	Corridor width (mm)	Grade ability (%)	Max speed (km/h)	Weight (kg)	Max thrust with cargo (N)
1	Jungheinrich	DFG 430s	Diesel	3300	3	4260	24	20.8	4376	18,100
2	Jungheinrich	TFG 430s	LPG	3300	3	4260	24	20.8	4376	18,100
3	Jungheinrich	DFG 430s	Diesel	7000	3	4260	24	20.8	4376	18,100
4	Jungheinrich	TFG 430s	LPG	7000	3	4260	24	20.8	4376	18,100
5	Jungheinrich	EFG 220	Battery	3000	2	3446	24	16	3382	12,300
6	Jungheinrich	EFG 220	Battery	6500	1.15	3446	24	16	3382	12,300
7	Jungheinrich	ETM 214	Battery	10,250	1.4	2757	14	14	2925	–
8	Jungheinrich	DFG 320s	Diesel	3300	2	3840	25	18.5	3270	10,550
9	Jungheinrich	TFG 320s	LPG	3300	2	3840	24	18	3250	12,650
10	Jungheinrich	DFG 320s	Diesel	6000	0.95	3840	25	18.5	3270	10,550
11	Jungheinrich	TFG 320s	LPG	6000	0.95	3840	24	18	3250	12,650
12	Jungheinrich	EFG 110	Battery	3000	1	3074	12,5	12	2570	4400
13	Jungheinrich	EFG 110	Battery	6000	0.8	3074	12.5	12	2570	4400
14	Jungheinrich	ETM 320 DZ	Battery	12,020	2	2883	10	14	3550	–
15	Jungheinrich	ETV 320 DZ	Battery	12,020	2	2804	10	14	3650	–
16	Jungheinrich	EFG 430	Battery	3100	3	4030	18	20	5100	14,000
17	Jungheinrich	EFG 430	Battery	7000	1.8	4030	18	20	5100	14,000
18	Jungheinrich	EFG 540	Battery	3100	4	4360	14	15	6600	14,600
19	Jungheinrich	EFG 550	Battery	3100	5	4360	12	15	7300	15,100
20	Jungheinrich	EFG 540	Battery	7175	3.2	4360	14	15	6600	14,600
21	Jungheinrich	EFG 550	Battery	7175	4	4360	12	15	7300	15,100

(continued)

Lp.	Manufacturer	Type	Drive	Lifting height (mm)	Capacity (t)	Corridor width (mm)	Grade ability (%)	Max speed (km/h)	Weight (kg)	Max thrust with cargo (N)
22	Jungheinrich	DFG 540	Diesel	3500	4	4619	25	25.3	6279	41,200
23	Jungheinrich	TFG 540	LPG	3500	4	4619	25	24.4	6279	41,200
24	Jungheinrich	DFG 540	Diesel	6775	3.38	4619	25	25.3	6279	41,200
25	Jungheinrich	TFG 540	LPG	6775	3.38	4619	25	24.4	6279	41,200
26	Jungheinrich	DFG 550	Diesel	3500	5	4769	23	24.8	7434	33,500
27	Jungheinrich	TFG 550	LPG	3500	5	4769	23	22.3	7434	33,500
28	Jungheinrich	DFG 550	Diesel	6675	3.85	4769	23	24.8	7434	33,500
29	Jungheinrich	TFG 550	LPG	6675	3.85	4769	23	22.3	7434	33,500
30	Jungheinrich	DFG 660	Diesel	3300	6	5206	28	23	10,060	45,700
31	Jungheinrich	DFG 670	Diesel	3300	7	5649	26	23	10,590	44,500
32	Jungheinrich	DFG 680	Diesel	3300	8	5799	24	23	11,000	44,300
33	Jungheinrich	DFG 690	Diesel	3300	9	5904	22	23	12,200	43,900
34	Jungheinrich	DFG 660	Diesel	6600	4.48	5206	28	23	10,060	45,700
35	Jungheinrich	DFG 670	Diesel	6600	6.09	5649	26	23	10,590	44,500
36	Jungheinrich	DFG 680	Diesel	6600	7.09	5799	24	23	11,000	44,300
37	Jungheinrich	DFG 690	Diesel	6600	7.82	5904	22	23	12,200	43,900
38	Still	RX60 - 20	Battery	3150	2	3648	18	20	3517	9663
39	Still	RX60 - 20	Battery	7915	0.75	3648	18	20	3517	9663
40	Still	RX60 - 30l	Battery	3020	3	4025	21.9	19	5097	17,070
41	Still	RX60 - 30l	Battery	7630	1.35	4025	21.9	19	5097	17,070
42	Still	RX60 - 40	Battery	2980	4	4408	15.5	19	6477	15,940

(continued)

Lp.	Manufacturer	Type	Drive	Lifting height (mm)	Capacity (t)	Corridor width (mm)	Grade ability (%)	Max speed (km/h)	Weight (kg)	Max thrust with cargo (N)
43	Still	RX60 - 50	Battery	2980	5	4408	13.2	19	7115	15,670
44	Still	RX60 - 40	Battery	7180	2.4	4408	15.5	19	6477	15,940
45	Still	RX60 - 50	Battery	7180	2.5	4408	13.2	19	7115	15,670
46	Still	R70 - 16	Diesel	3330	1.6	3695	25	21	2640	12,000
47	Still	R70 - 16t	LPG	3330	1.6	3695	25	21	2640	12,000
48	Still	R70 - 20c	Diesel	3330	2	3817	25	21	3090	12,000
49	Still	R70 - 20t	LPG	3330	2	3817	25	21	3090	12,000
50	Still	R70 - 16	Diesel	8020	0.6	3695	25	21	2640	12,000
51	Still	R70 - 16t	LPG	8020	0.6	3695	25	21	2640	12,000
52	Still	R70 - 20c	Diesel	8065	0.6	3817	25	21	3090	12,000
53	Still	R70 - 20t	LPG	8065	0.6	3817	25	21	3090	12,000
54	Still	R70 - 40	Diesel	3180	4	4618	24	21	5800	22,230
55	Still	R70 - 40t	LPG	3180	4	4618	24	21	5800	22,230
56	Still	R70 - 50	Diesel	3180	5	4710	20	21	6395	22,110
57	Still	R70 - 50t	LPG	3180	5	4710	20	21	6395	22,110
58	Still	R70 - 40	Diesel	7180	2.4	4618	24	21	5800	22,230
59	Still	R70 - 40t	LPG	7180	2.4	4618	24	21	5800	22,230
60	Still	R70 - 50	Diesel	7180	3	4710	20	21	6395	22,110
61	Still	R70 - 50t	LPG	7180	3	4710	20	21	6395	22,110
62	Still	R70 - 60	Diesel	3500	6	4696	31	24	8824	45,230
63	Still	R70 - 70	Diesel	3500	7	4818	24	24	10,560	45,230
64	Still	R70 - 80	Diesel	3500	8	5218	24	24	10,667	45,230
65	Still	RX20 - 15	Battery	3230	1.5	3328	21.2	16	2824	9260

(continued)

Lp.	Manufacturer	Type	Drive	Lifting height (mm)	Capacity (t)	Corridor width (mm)	Grade ability (%)	Max speed (km/h)	Weight (kg)	Max thrust with cargo (N)
66	Still	RX20 - 15	Battery	7870	0.45	3328	21.2	16	2824	9260
67	Linde	E12	Battery	3110	1.2	3164	15.6	12.5	2680	6450
68	Linde	E20	Battery	3150	2	3599	16	15.5	3660	9220
69	Linde	E12	Battery	5475	0.6	3164	15.6	12.5	2680	6450
70	Linde	E20	Battery	6765	0.7	3599	16	15.5	3660	9220
71	Linde	E30	Battery	3050	3	3872	14	15	4845	11,702
72	Linde	E30	Battery	6605	1.1	3872	14	15	4845	11,702
73	Linde	E40	Battery	3250	4	4440	14	14	6870	14,200
74	Linde	E40	Battery	5550	2.7	4400	14	14	6870	14,200
75	Linde	H14 d	Diesel	3110	1.4	3770	35	20	2590	12,900
76	Linde	H14	LPG	3110	1.4	3770	35	20	2570	12,900
77	Linde	H20 d	Diesel	3110	2	3895	27	20	3060	12,900
78	Linde	H20	LPG	3110	2	3895	27	20	3040	12,900
79	Linde	H14 d	Diesel	5475	0.7	3770	35	20	2590	12,900
80	Linde	H14	LPG	5475	0.7	3770	35	20	2570	12,900
81	Linde	H20 d	Diesel	5475	1.2	3895	27	20	3060	12,900
82	Linde	H20	LPG	5475	1.2	3895	27	20	3040	12,900
83	Linde	H30 d	Diesel	4705	3	4289	27	22	4220	19,790
84	Linde	H30	LPG	4705	3	4289	26	22	4200	19,790
85	Linde	H30 d	Diesel	6465	1	4289	27	22	4220	19,790
86	Linde	H30	LPG	6465	1	4289	26	22	4200	19,790
87	Linde	H40 d	Diesel	4675	4	4555	29	21	5745	28,541
88	Linde	H40	LPG	4675	4	4555	28	21	5745	28,541

(continued)

Lp.	Manufacturer	Type	Drive	Lifting height (mm)	Capacity (t)	Corridor width (mm)	Grade ability (%)	Max speed (km/h)	Weight (kg)	Max thrust with cargo (N)
89	Linde	H50 d	Diesel	4525	5	4680	21	24	6580	25,285
90	Linde	H50	LPG	4525	5	4680	20	24	6580	25,285
91	Linde	H40 d	Diesel	6315	2.1	4555	29	21	5745	28,541
92	Linde	H40	LPG	6315	2.1	4555	28	21	5745	28,541
93	Linde	H50 d	Diesel	6315	2.9	4680	21	24	6580	25,285
94	Linde	H50	LPG	6315	2.9	4680	20	24	6580	25,285
95	Linde	H60 D	Diesel	3550	6	5090	22	23	10,160	37,564
96	Linde	H70 D	Diesel	3150	7	5100	23	23	10,400	44,968
97	Linde	H80 D	Diesel	3150	8	5100	20	23	12,520	44,968
98	Linde	H60 D	Diesel	6050	4	5090	22	23	10,160	37,564
99	Linde	H70 D	Diesel	5650	5	5100	23	23	10,400	44,968
100	Linde	H80 D	Diesel	5650	5.5	5100	20	23	12,520	44,968
101	Linde	R16 X	Battery	6355	1.6	2761	10	14	3810	–
102	Linde	R16 X	Battery	11,455	1.6	2761	10	14	3810	–

Appendix 2
Forklift Results

The study was conducted in order to verify the computational model for the evaluation of energy intensity of a logistic storage system. The research was conducted in terms of energy intensity and the drag coefficient.

The study was conducted in the actual operating conditions in the facility (floor dust binding, horizontal (no ramps), covered with a non-slip layer) for the forklift STILL RX60-25; this is an electricity-driven forklift, without energy recovery. The energy intensity was measured using the measuring apparatus while traversing a route from "A" to "B" and then back to "A". The transport process is divided into stages: 1–2, 2–3, 3–4, 4–5, 5–6, 6–7, 7–8, 8–9, up/down power.

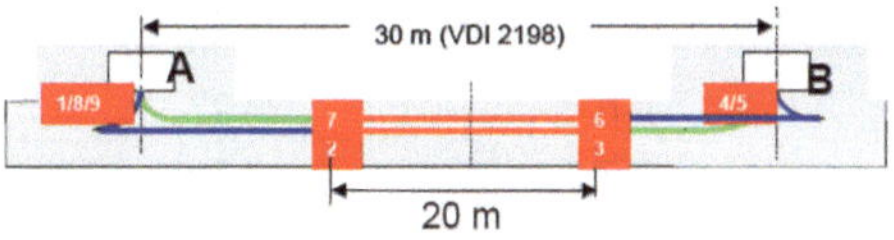

Fig. A.1 Layout of the forklift route for comparative studies

The study was conducted for eight types of tires: Continental (measurement-1), Gumasol (measurement-2), Marangoni (measurement-3), Watts (measurement-4), Bergougnan (measurement-5), Bergougnan (measurement-6), Trelleborg (measurement-7), Solideal (measurement-8).

P. Zajac, *Evaluation Method of Energy Consumption in Logistic Warehouse Systems*, EcoProduction, DOI 10.1007/978-3-319-22044-4

(Measurement-1)			
Measurement number	Section	Time (s)	Energy (J)
1	1–2	16.59	103,101
	2–3	8.55	60,828
	3–6	24.68	114,128
	6–7	8.50	60,366
	7–9	52.17	358,361
	Total	110.49	696,784
2	1–2	17.41	105,884
	2–3	8.50	60,901
	3–6	23.81	110,624
	6–7	8.50	60,811
	7–9	55.35	334,448
	Total	113.57	672,668
	1–2	15.97	101,660
	2–3	8.50	60,314
	3–6	23.86	112,736
	6–7	8.50	60,353
	7–9	52.38	317,789
	Total	109.21	652,852
	1–2	15.56	100,670
	2–3	8.50	60,052
	3–6	24.27	110,560
	6–7	8.50	60,469
	7–9	52.74	353,795
	Total	109.57	685,546
	1–2	15.82	102,254
	2–3	8.50	59,171
	3–6	22.53	110,070
	6–7	8.50	59,985
	7–9	56.83	318,332
	Total	112.18	649,812
Mean energy			675,250

(Measurement-2)			
Measurement number	Section	Time (s)	Energy (J)
1	1–2	16.54	103,396
	2–3	8.50	60,456
	3–6	27.60	124,603
	6–7	8.50	59,963
	7–9	55.60	354,060
	Total	116.74	702,478
2	1–2	16.64	104,925
	2–3	8.50	61,337
	3–6	28.36	126,730
	6–7	8.50	61,414
	7–9	58.78	357,662
	Total	120.78	712,068
3	1–2	16.95	104,163
	2–3	8.50	59,627
	3–6	28.31	126,961
	6–7	8.50	61,465
	7–9	55.65	345,718
	Total	117.91	697,934
4	1–2	16.38	102,469
	2–3	8.50	61,224
	3–6	26.37	116,726
	6–7	8.50	60,245
	7–9	54.53	348,224
	Total	114.28	688,888
	1–2	17.10	103,838
	2–3	8.50	60,724
	3–6	27.14	119,690
	6–7	8.50	60,735
	7–9	54.17	342,746
	Total	115.41	687,733
Mean energy			698,675

(Measurement-3)			
Measurement number	Section	Time (s)	Energy (J)
1	1–2	17.82	96,464
	2–3	8.40	45,562
	3–6	13.09	96,751
	6–7	8.40	45,782
	7–9	55.40	339,492
	Total	113.11	624,051
2	1–2	17.82	94,478
	2–3	8.40	44,886
	3–6	23.60	98,518
	6–7	8.40	46,127
	7–9	59.60	357,162
	Total	117.82	641,171
3	1–2	18.48	97,943
	2–3	8.40	45,269
	3–6	24.27	99,112
	6–7	8.40	45,334
	7–9	57.70	326,860
	Total	117.25	614,518
4	1–2	17.41	95,187
	2–3	8.35	44,360
	3–6	23.50	99,735
	6–7	8.35	44,875
	7–9	57.75	321,673
	Total	115.36	605,830
5	1–2	15.72	89,204
	2–3	8.40	45,272
	3–6	24.52	100,422
	6–7	8.40	43,955
	7–9	61.70	361,152
	Total	118.74	640,005
Mean energy			619,452

(Measurement-4)			
Measurement number	Section	Time (s)	Energy (J)
1	1–2	16.79	96,276
	2–3	8.50	49,555
	3–6	23.60	105,736
	6–7	8.50	50,427
	7–9	49.72	320,292
	Total	107.11	622,286
2	1–2	18.07	103,290
	2–3	8.55	51,860
	3–6	24.22	103,537
	6–7	8.50	51,478
	7–9	54.84	324,231
	Total	114.18	634,396
3	1–2	16.90	96,741
	2–3	8.50	50,708
	3–6	24.83	109,787
	6–7	8.50	50,372
	7–9	56.27	318,737
	Total	115.00	626,345
4	1–2	17.61	95,569
	2–3	8.55	50,530
	3–6	24.78	103,985
	6–7	8.50	50,404
	7–9	56.88	364,864
	Total	116.32	665,352
5	1–2	18.28	98,642
	2–3	8.50	49,451
	3–6	24.73	101,496
	6–7	8.50	49,728
	7–9	52.68	328,516
	Total	112.69	627,833
Mean energy			642,072

(Measurement-5)			
Measurement number	Section	Time (s)	Energy (J)
1	1–2	16.18	100,547
	2–3	8.60	54,319
	3–6	21.25	102,544
	6–7	8.55	52,974
	7–9	54.63	352,849
	Total	109.21	663,233
	1–2	17.46	100,842
	2–3	8.55	54,874
	3–6	22.84	105,271
	6–7	8.55	54,562
	7–9	55.81	333,195
	Total	113.20	648,744
	1–2	15.67	92,816
	2–3	8.55	53,519
	3–6	24.47	108,509
	6–7	8.55	54,468
	7–9	52.79	346,703
	Total	110.03	656,015
	1–2	16.18	96,903
	2–3	8.60	54,233
	3–6	23.71	111,235
	6–7	8.50	53,791
	7–9	57.91	365,491
	Total	114.90	681,653
	1–2	15.82	95,801
	2–3	8.55	53,777
	3–6	22.37	103,553
	6–7	8.55	53,810
	7–9	53.66	339,332
	Total	108.95	646,273
Mean energy			655,863

(Measurement-6)			
Measurement number	Section	Time (s)	Energy (J)
1	1–2	18.48	107,820
	2–3	8.60	63,134
	3–6	22.68	112,170
	6–7	8.55	63,492
	7–9	52.28	351,936
	Total	110.59	698,552
2	1–2	16.23	104,018
	2–3	8.55	62,551
	3–6	24.52	115,459
	6–7	8.55	63,831
	7–9	57.91	340,393
	Total	115.76	686,252
3	1–2	15.97	99,851
	2–3	8.55	62,998
	3–6	26.78	118,420
	6–7	8.50	62,223
	7–9	55.76	350,096
	Total	115.56	693,588
	1–2	16.49	101,682
	2–3	8.55	62,752
	3–6	26.21	118,254
	6–7	8.55	62,439
	7–9	55.55	360,835
	Total	115.35	705,962
	1–2	15.82	102,295
	2–3	8.55	63,082
	3–6	26.78	119,674
	6–7	8.50	62,195
	7–9	55.65	336,799
	Total	115.30	684,045
Mean energy			690,683

(Measurement-7)			
Measurement number	Section	Time (s)	Energy (J)
1	1–2	16.54	105,726
	2–3	8.55	60,211
	3–6	23.40	112,573
	6–7	8.55	61,948
	7–9	51.10	332,625
	Total	108.14	673,083
2	1–2	16.13	100,526
	2–3	8.55	61,544
	3–6	24.12	110,232
	6–7	8.55	59,592
	7–9	55.14	351,781
	Total	112.49	683,675
3	1–2	17.61	104,924
	2–3	8.55	60,770
	3–6	24.68	112,211
	6–7	8.55	62,176
	7–9	54.17	351,446
	Total	113.56	691,527
4	1–2	18.43	107,807
	2–3	8.50	59,830
	3–6	25.86	113,740
	6–7	8.55	61,041
	7–9	58.37	340,208
	Total	119.71	682,626
5	1–2	18.64	106,671
	2–3	8.55	60,837
	3–6	25.80	117,798
	6–7	8.55	60,806
	7–9	60.52	344,817
	Total	122.06	690,929
Mean energy			679,725

(Measurement-8)			
Measurement number	Section	Time (s)	Energy (J)
1	1–2	18.12	103,162
	2–3	8.50	53,946
	3–6	27.80	117,608
	6–7	8.50	55,494
	7–9	56.93	349,091
	Total	119.85	679,301
2	1–2	18.89	103,801
	2–3	8.50	56,311
	3–6	25.91	111,402
	6–7	8.50	54,922
	7–9	54.32	326,409
	Total	116.12	652,845
3	1–2	18.07	101,330
	2–3	8.50	54,562
	3–6	28.01	114,366
	6–7	8.50	54,604
	7–9	49.97	322,376
	Total	113.05	647,238
4	1–2	16.28	96,756
	2–3	8.50	54,201
	3–6	26.98	109,461
	6–7	8.50	54,903
	7–9	55.35	338,243
	Total	115.61	653,564
5	1–2	17.05	97,698
	2–3	8.55	54,564
	3–6	27.65	113,061
	6–7	8.50	54,924
	7–9	58.93	348,693
	Total	120.68	668,940
Mean energy			667,398

The difference between experimental results and the results of the computational model for energy intensity is no more than 2 %

References

1. Adamczyk, E., Jucha, J., Miller, S.: Theory of Mechanisms and Machines. Publishing PWN, Warsaw (1977)
2. Agou, M., Nishi, T., Konishi, M.I.: A dynamic optimization model for storage yard logistic systems. Proceedings of the SICE Annual Conference in Okayama, SICE 2005, pp. 3254–3259 (2005)
3. Alvarez-Valdes, R., Parrefto, F., Tamarit, J.M.: A tabu search algorithm for the pallet loading problem. OR Spectrum **27**(1), 43–61 (2005)
4. Andrzejewski, R.: The Dynamics of Pneumatic Running Wheel. Scientific and Technical Publishing, Warsaw (2010)
5. Ashayeri, J., Gelders, L., Wassenhove, L.V.: A microcomputer-based optimization model for the design of automated warehouses. Int. J. Prod. Res. **23**(4), 825–839 (1995)
6. Ashayeri, J., Heuts, R.M., Yalkenburg, M.W.T., Yeraart, H.C., Wilhelm, M.R.: A geometrical approach to computing expected cycle times for zone-based storage layouts in AS/RS. Int. J. Prod. Res. **40**(17), 4467–4483 (2002)
7. Azadivar, F.: Maximizing of the through put of a computerized automated warehousing system under system constraints. Int. J. Prod. Res. **24**(3), 551–566 (1986)
8. Azadivar, F.: Optimum allocation of resources between the random access and rack storage spaces in an automated warehousing system. Int. J. Prod. Res. **27**(1), 119–131 (1998)
9. Barcik, R., Bukowski, L., Michlowicz, E., Nowicka-Skowron, M., Sawik, T., Bukowski, L.: Problems of Logistics Information Processing in Integrated Production Systems. Publishing "TEXT", Krakow (2004)
10. Barcik, R., Bukowski, L., Michlowicz, E., Nowicka-Skowron, M., Sawik, T., Fialkowski, J.: Spaces Buffer and Their Technological Equipment in the Optimization of Logistics Systems —Identification of the Problem with the Analysis of the Particular Case. Publishing "TEXT", Krakow (2004)
11. Barcik, R., Bukowski, L., Michlowicz, E., Nowicka-Skowron, M., Sawik, T., Ratkiewicz, A. (eds.): Application of Hellwig to Choose Technological Parameters Influencing the Cycle Time Picking. Publishing "TEXT", Krakow (2004)
12. Bartholdi, J.J., Hackman, S.T.: Warehouse and Distribution Science. www.warehouse-sciene.com, Atlanta USA (2010)
13. Bartholdi, J.J., Hackamn, S.T.: Warehouse & Distribution Science. The Supply Chain and Logistics Institute School of Industrial and System Engineering, Atlanta USA (2010)
14. Bartholdi, J.J., Eisenstein, D.D., Foley, R.D.: Performance of bucket brigades when work is stochastic. Oper. Res. **49**(5), 710–719 (2000)
15. Basham, D.L., Wright, J.W., Ferguson, K.T., Moy, G.W.: Design: Covered Storage. Department of Defense USA (1982)
16. Bassan, Y., Roli, Y., Rosenblatt, M.J.: Internal layout design of a warehouse. AIIE Trans. **12** (4), 317–322 (1990)
17. BEA: 2012. http://bea.gov

P. Zajac, *Evaluation Method of Energy Consumption in Logistic Warehouse Systems*, EcoProduction, DOI 10.1007/978-3-319-22044-4

18. Behnamian, J., Eghtedari, B.: Storage System Layout. Contributions to Management Science, Facility Location (2001)
19. Bejan, A., Tsatsaronis, G., Moran, M.: Thermal Design and Optimization. Wiley, New York (1996)
20. Yincent, B.G., Kung, H.-C., Leblanc, J.A., Joan, M.A.: Sprinkler protection for warehouse storage of flammable in small metal containers. J. Fire Prot. Eng. **9**, 14–35 (1998)
21. Benwell, M.: Scheduling stocks and storage space in a volatile market. Logistics Inf. Manage. **9**(4), 18–23 (1996)
22. Berry, J.R.: Elements of warehouse layout. Int. J. Prod. Res. **7**(2), 105–121 (1998)
23. Bhaskaran, K., Malmborg, C.J.: Modelling the service process in a multi-address warehousing system. Appl. Math. Model. **13**(7), 386–396 (2003)
24. Biegus, A.: Steel Buildings Indoor. Publishing Arkady, Warsaw (2009)
25. Bloss, R.: Pallet conveyor now smarter and slimmer. Assembly Autom. **25**(2), 106–107 (2005)
26. Bodziony, P., Furmanik, K.: The selection of technological cars in the mines of rock raw materials. Ind. Transp. **1**(31) (2008)
27. Bozer, Y.A., White, J.A.: A generalized design and performance analysis models for end-of-aisle order-picking systems. IIE Trans. **28**, 271–280 (2006)
28. Bozer, Y.A., White, J.A.: Design and performance models for end-of-aisle order picking systems. Manage. Sci. **36**(7), 852–866 (2003)
29. Bozer, Y.A., White, J.A.: Travel-time models for automated storage/retrieval systems. IIE Trans. **16**(4), 329–338 (2003)
30. Bruhns, H., Steadman, P., Herring, H.: A database for modeling energy use in the non-domestic building stock of England and Wales. Appl. Energy **66**, 277–297 (2000). www.elsevier.com/locate/apenergy
31. Brynzer, H., Johansson, M.I.: Design and performance of kitting and order picking systems. Int. J. Prod. Econ. **41**, 115–125 (2005)
32. Bujak, A., Zajac, P.: Can the increasing of energy consumption of information interchange be a factor that reduces the total energy consumption of a logistic warehouse system? In: Telematics in the Transport Environment: 12th International Conference on Transport Systems Telematics, pp. 199–210, TST 2012, Springer, cop. 2012
33. Burkard, R.E., Fruhwirth, B., Rote, G.: Vehicle routing in an automated warehouse, analysis and optimization. Ann. Oper. Res. **57**, 29–44 (2006)
34. Button, G., Kenneth, J., Hensher, D.A.: Handbook of Transport Systems and Traffic Control, vol. 3. Emerald, Inc., USA (2009)
35. Caron, F., Marchet, G., Perego, A.: Optimal layout in low-level picker-to-part systems. Int. J. Prod. Res. **38**(1), 101–117 (2010)
36. Caron, F., Marchet, G., Perego, A.: Routing policies and storage policies in picker-to- part systems. Int. J. Prod. Res. **36**(3), 713–732 (2008)
37. Chang, D.T., Wen, U.P., Lin, J.T.: The impact of acceleration/deceleration on travel- time models for automated storage/retrieval systems. IIE Trans. **27**, 108–111 (2005)
38. Chang, D.T., Wen, U.P.: The impact of rack configuration on the speed profile of the storage and retrieval machine. IIE Trans. **29**, 525–531 (2007)
39. Chew, E.P., Tang, L.C.: Travel time analysis for general item location assignment in a rectangular warehouse. Eur. J. Oper. Res. **112**, 582–597 (2009)
40. Chew, W.M.: An analysis of automated storage and retrieval systems in manufacturing assembly lines. IIE Trans. **18**(2), 204–214 (2003)
41. Christopher, T.: Pallet stacking racks work anywhere. Mater. Handling Eng. **49**(11), 71–74 (1994)
42. Cichonski, T., Jezusek, M.: Modern warehouse of STIHL. Logistics Mag. **2**, 41–44 (1999)
43. Cieslak, M. (ed.): Prognozowanie Gospodarcze. Wydawnictwo Naukowe PWN, Warsaw (2001)

44. Connelly, L., Koshland, C.P.: Exergy and industrial ecology, part 1: an exergy-based definition of consumption and a thermodynamics interpretation of ecosystem evolution. Exergy Int. J. **1**(3), 146–165 (2001)
45. Corey, G.P.: Energy storage solutions for premium power. Aerosp. Electron. Syst. Warehouse IEEE **11**(6), 41–44 (1996)
46. Cormier, G., Gunn, E.A.: A review of warehouse models. Eur. J. Oper. Res. **58**, 3–13 (1992)
47. Cormier, G., Gunn, E.A.: Modelling and analysis for capacity expansion planning in warehousing. J. Oper. Res. Soc. **50**(1), 52–59 (2009)
48. Cormier, G., Gunn, E.A.: On coordinating warehouse sizing, leasing and inventory policy. IIE Trans. **28**, 149–154 (2006)
49. Cormier, G., Kersey, D.F.: Conceptual design of a warehouse for just-in-time operations in a bakery. Comput. Ind. Eng. **29**(1–4), 361–365 (2005)
50. Cormier, G.: Operational research methods for efficient warehousing. In: Langevin, A., Riopel, D. (eds.) Logistics Systems: Design and Optimization, pp. 93–122. Springer, Berlin (2005)
51. Cornelissen, R.L.: Thermodynamics and sustainable development. The use of exergy analysis and the reduction of reversibility. Ph.D. thesis, University of Twente, Holland (1997)
52. Cox, B.: Determining economic levels of automation by using a hierarchy of productivity ratios techniques. Proceedings of 7th International Conference on Automation in Warehousing (1986)
53. Cyganiuk, J., Slonski, E.: Portal robots in logistic systems of completation of cumulative pallet cargo systems. J. Transdisciplinary Syst. Sci. **8**(l), 16–23 (2003)
54. Daskin, M.S., Snyder, L.V., Berger, R.T.: Facility location in supply chain design. In: Langevin, A., Riopel, D. (eds.) Logistics Systems: Design and Optimization, pp. 39–65. Springer Science Business Media Inc, New York (2005)
55. Dekker, R., Koster, R., Roodbergen, K.J., van Kalleveen, H.: Improving order-picking response time at Ankor's warehouse. Interfaces **34**(4), 303–313 (2004)
56. d'Hont, S.: Smart pallet system improves warehouse productivity. Sens. Rev. **16**(4), 21–24 (1996)
57. Drobinski, W., Tali, M.: The Drive and Electrical Appliances. Wroclaw University of Technology Press, Wroclaw (1980)
58. Duditza, F.: Kardangelangetriebe und ihre Anwendungen. VDI-Verlag, Dusseldorf (1973)
59. Dudzinski, Z., Kizyn, M.: Warehouse Guide. Polish Economic Publishing House, Warsaw (2008)
60. Durski, W., Redmer, A.: Monitoring the status of cargo during transport. Logistics Magazine 3/2008, Poznan (2008)
61. Escudero, A.T., Coppola, L., Marschoff, C.M.: Technology substitution in the energy market: the logistic approach revisited. Energy Convers. Manage. **5**, 415–441 (1997)
62. Eynan, A., Rosenblatt, M.J.: Establishing zones in single-command class-based rectangular AS/RS. Transactions **26**(1), 38–46 (1994)
63. Falk, N.K.: Warehouse lighting—it costs or pays an energy management approach. Industrial Engineering Conference, American Institute of Industrial Engineers, pp. 119–124 (1993)
64. Fialkowski, J.: Designing High Rack Warehouse. Publishing ARKADY, Warsaw (1983)
65. Fialkowski, J.: Warehousing Technology. Warsaw University of Technology Publishing House, Warsaw (1995)
66. Fice, M.: Capacitor Energy Storage in Diesel-Electric Drive. Cracow University of Technology Publishing House, Krakow (2008)
67. Fogel, M., Burkhart, N., Ren, H., Schift, J., Mens, M., Goldberg, K.: Automated tracking of pallets in warehouses: Beacon layout and asymmetric ultrasound observation models. Proceedings of the 3rd IEEE International Conference on Automation Science and Engineering, IEEE CASE 2007, pp. 678–685 (2007)

68. Foley, R., Frazelle, E.H., Park, B.C.: Throughput bounds for mini load automated storage/retrieval systems. IIE Trans. **34**(10), 915–920 (2002)
69. Foley, R., Frazelle, E.H.: Analytical results for mini load throughput and the distribution of dual command travel time. IIE Trans. **23**(3), 273–281 (1999)
70. Francis, R.L.: On some problems of rectangular warehouse design and layout. J. Ind. Eng. **18**, 595–604 (1997)
71. From the palletizer to the warehouse. Source: Glass International **22**(3), 34 May–Jun 1999
72. Galuszka, Z., Szybka, J.: The Problem of Optimizing the Decision When Mixed Criteria of Judgment. Scientific Papers AGH, Krakow (1977)
73. Gamberini, R., Grassi, A., Mora, C., Rimini, B.: An innovative approach for optimizing warehouse capacity utilization. Int. J. Logistics Res. Appl.: A Lead. J. Supply Chain Manage., Jan 1, 1469, **11**(2), 137–165 (2008)
74. Goetschalckx, M., Ratliff, H.D.: Optimal lane depths for single and multiple products in block stacking storage systems. IIE Trans. **23**(3), 245–258 (1991)
75. Goetschalckx, M., Ratliff, H.D.: Shared storage policies based on the duration stay of unit loads. Manage. Sci. **36**(9), 1120–1132 (1990)
76. Goh, M., Jihong, O., Chung-Piaw, T.: Warehouse sizing to minimize inventory and storage costs. Naval Res. Logistics **48**(4), 299–312 (2001)
77. Gorski, K.: Automatic storage of flat products of wood. Logistics Mag. **4**, 44–45 (1999)
78. Graves, S.C., Hausman, W.H., Schwarz, L.B.: Storage-retrieval interleaving in automatic warehousing systems. Manage. Sci. **23**(9), 935–945 (1977)
79. Gray, A.E., Karmarkar, S., Seidmann, A.: Design and operation of an order—consolidation warehouse: models and applications. Eur. J. Oper. Res. **58**, 14–36 (1992)
80. Gronowicz, J.: Energy Management in Land Transport. Publisher Technical University of Poznan, Poznan (2006)
81. Gu, J.X., Goetschalckx, M., McGinnis, L.F.: Research on warehouse operation: a comprehensive review. Eur. J. Oper. Res. **177**(1), 1–21 (2007)
82. Guang-Zhao, C., Lin-Sha, L., Zhen-Dong, H., Li-Na, Y., Cun-Xiang, Y., Bu-Yi, H., Zhi-Hong, H.: A robust autonomous mobile forklift pallet recognition. 2nd International Asia Conference on Informatics in Control, Automation and Robotics (CAR), Jan 1, 1109, vol. 3 (10), pp. 286–290 (2010)
83. Gudehus, T.: Grundlagen der Kommissioniertechnik, Dynamik der Warenverteil und Lagersysteme. Verlag W. Girardet, Essen (1973)
84. Gudehus, T.: "Logistik 1" Grundlagen, Verfahren, und Strategien. Springer, Berlin (2006)
85. Gudehus, T.: "Logistik 2" Netzwerke, Systeme und Lieferketten. Springer, Berlin (2006)
86. Gupta, B.D., Rao, S.S.: Automated optimum design of refrigerated warehouses. ASME PAP, WA/DE-11, Issue 1978
87. Ha, J.-W., Hwang, H.: Class-based storage assignment policy in carousel system. Comput. Ind. Eng. **26**(3), 489–499 (1994)
88. Hackman, S.T., Frazelle, E.H., Griffin, P.M., Griffin, S.O., Vlasta, D.A.: Benchmarking warehouse and distribution operations: an input-output approach. J. Prod. Anal. **16**, 79–100 (2001)
89. Hali, R.W.: Distance approximation for routing manual pickers in a warehouse. IIE Trans. **25** (4), 76–87 (1993)
90. Hamrol, A., Mantura, W.: "Quality Management" Theory and Practice. Publishing PWN, Warsaw (2005)
91. Han, M.H., McGinnis, L.F.: Carousel Application for Work-in-process: Modelling and Analysis. Material Handling Research Center, Georgia Institute of Technology, Atlanta, Georgia (1996)
92. Han, M.H., McGinnis, L.F., Shieh, J.S., White, J.A.: On sequencing retrievals in an automated storage/retrieval system. IIE Trans. **19**(1), 56–66 (1997)
93. Han, M.H., McGinnis, L.F., White, J.A.: Analysis of rotary rack operation. Mater. Flow **4**, 283–293 (1994)

94. Harrom, H.A.: Energy conservation—electric forklift hydraulic systems. SAE Preprints **34**, 750–804, Jan 1, 1995
95. Hausman, W.H., Schwarz, L.B., Graves, S.C: Optimal storage assignment in automatic warehousing systems. Manage. Sci. **22**(6), 629–638 (2004)
96. Heragu, S.S., Du, L., Mantel, R.J., Schuur, P.C.: Mathematical model for warehouse design and product allocation. Int. J. Prod. Res. **43**(2), 327–338 (2005)
97. Ho, S.S., Sarma, S.: The fragmented warehouse: location assignment for unit-load picking. Ind. Eng. Eng. Manage. **2008**, 1159–1163, Jan 1, 2008
98. Hoefkens, L.J., Orwin, O.J.B., Shield, C., Read, N.K.: Higher stacking. An examination of the special conditions associated with the high stacking of pallets by means of fork lift trucks. Prod. Eng. **43**(9), Jan 1, 1994
99. Hogan, B.J.: Carousel's walking beams move pallets, shotpins locate them. Des. News (Boston) **42**(16), 132–134 (2006)
100. Hompel, M., Schmidt, T.: "Management of Warehouse Systems Intralogistik', Warehouse Management. Springer, Berlin (2007)
101. Hompel, M., Siebel, L.: Logistik und e-commerce: Koncepte fur Ballungszentren. Praxiswissen, Deutschland (2001)
102. Hung, M.S., Fisk, C.J.: Economic sizing of warehouses—a linear programming approach. Comput. Oper. Res. **11**(1), 13–18 (1984)
103. Hur, S., Lee, Y.H., Lim, S.Y., Lee, M.H.: A performance estimating model for AS/ RS by M/G/1 queuing system. Comput. Ind. Eng. **46**, 233–241 (2004)
104. Hwang, H., Lee, S.B.: Travel-time models considering the operating characteristics of the storage and retrieval machine. Int. J. Prod. Res. **28**(10), 1779–1789 (1990)
105. Hwang, H., Ha, J.-W.: Cycle time models for single/double carousel system. Int. J. Prod. Econ. **25**, 129–140 (1991)
106. Hwang, H., Kim, C.-S., Ko, K.-H.: Performance analysis of carousel systems with double shuttle. Comput. Ind. Eng. **36**, 473–485 (1999)
107. Hwang, H., Ko, C.S.: A study on multi-aisle system served by a single storage/retrieval machine. Int. J. Prod. Res. **26**(11), 1727–1737 (1988)
108. Hwang, H., Oh, Y.H., Lee, Y.K.: An evaluation of routing policies for order-picking operations in low-level picker-to-part system. Int. J. Prod. Res. **42**(18), 3873–3889 (2004)
109. Hwang, H., Song, J.Y.: Sequencing picking operations and travel time models for man-on-board storage and retrieval warehousing system. Int. J. Prod. Econ. **29**, 75–88 (1993)
110. Hwang, H., Song, Y.-K., Kim, K.-H.: The impacts of acceleration/deceleration on travel time models for carousel systems. Comput. Ind. Eng. **46**, 253–265 (2004)
111. Ignasiak, E., Borucki, W., Marcinkowski, J., Sikora, W. (eds.): Operational Research. Polish Economic Publishing House, Warsaw (2001)
112. Ito, T., Abadi, J., Mousavi, S.M.: Agent-based material handling and inventory planning in warehouse. J. Intell. Manuf. **13**(3), 201–210 (2002)
113. Jakubowski, L.: Technology Loading Work. Warsaw University of Technology Publishing House, Warsaw (2003)
114. Jara, J.A.: The impact of polyurethane tires on the vehicle motion energy consumption. Promoter Dudzinski, P., Preprints Series Report no. PRE 014/01. Wroclaw University of Technology, Wroclaw (2001)
115. Jaskiewicz, Z.: The Design of Propulsion Systems of Motor Vehicles. Publishing WKL, Warsaw (1982)
116. Jawis, J.M., McDowell, E.D.: Optimal product layout in an order picking warehouse. IIE Trans. **23**(1), 93–102 (1991)
117. Jedrzejczyk, Z., Kukula, K.: Operational Research in Examples and Tasks. Publishing PWN, Warsaw (1997)
118. Johnson, M.E., Lofgren, T.: Model decomposition speeds distribution center design. Interfaces **24**(5), 95–106 (1994)

119. Johnson, M.E., Meller, R.D.: Performance analysis of split-case sorting systems. Manuf. Serv. Oper. Manage. **4**(4), 258–274 (2002)
120. Kallina, C., Lynn, J.: Application of the cube-per-order index rule for stock location in a distribution warehouse. Interfaces **7**(1), 37–46 (1976)
121. Kallrath, J.: Online Storage Systems and Transportation Problems with Applications. Springer, ITWM Germany (2005)
122. Karasawa, Y., Nakayama, H., Dohi, S.: Trade-off analysis for optimal design of automated warehouses. Int. J. Syst. Sci. **11**(5), 567–576 (1980)
123. Karthik, A.: Integrated Analytical Performance Evaluation Models of Warehouses. Ph.D., Oklahoma State University (2009)
124. Keisuke, T., Sato, A., Osamu, N., Masayuki, M., Shinichi, K.: High-efficiency technology for 3-wheel electric forklift truck. Tech. Rev. Mitsubishi Heavy Ind. **37**(1), 14–18 (2000)
125. Kim, B.-I., Graves, R.J., Heragu, S.S., Onge, A.S.: Intelligent agent modeling of an industrial warehousing problem. IIE Trans. **34**(7), 601–612 (2002)
126. Kim, J., Seidmann, A.: A framework for the exact evaluation of expected cycle times in automated storage systems with full-turnover item allocation and random service requests. Comput. Ind. Eng. **18**(4), 601–612 (1990)
127. Kiranmai, B.: Improving reliability, energy-efficiency and security of storage systems and real-time systems, pp. 145–159. Ph.D., Auburn University (2009)
128. Klimczak, B.: Microeconomic. Publisher University of Economics, Oskar Lange in Wroclaw, Wroclaw (1995)
129. Knill, B.: How palletizers stack up. Mod. Mater. Handling **60**(6), 35–37 (2005)
130. Koh, S.G., Kim, B.S., Kim, B.N.: Travel time model for the warehousing system with a tower crane S/R machine. Comput. Ind. Eng. **43**(3), 495–507 (2002)
131. Kondratowicz, L.: EDI in Transport Logistics. Publications of the University of Gdansk, Gdansk (1999)
132. Kondratowicz, L.: Electronic Data Exchange Trading and Maritime Transport. Publications of the University of Gdansk, Gdansk (1995)
133. Korzen, Z. (ed.): Logistics in the Transport of Goods, NAVIGATOR. Wroclaw University of Technology Press, Wroclaw (1998)
134. Korzen, Z.: Ecologistics. Publishing ILiM, Poznan (2001)
135. Korzen, Z.: Logistics Handling Systems and Storage, vol. 1. Publishing ILiM, Poznan (1997)
136. Korzen, Z.: Logistics Handling Systems and Storage, vol. 2. Publishing ILiM, Poznan (1999)
137. Korzen, Z.: "Fundamentals of Logistics" Work Project TEMPUS JEP-03238 No. 1/1995, ICandEM. Wroclaw University of Technology, Wroclaw (1995)
138. Koster, R.: Performance approximation of pick-to-belt order picking systems. Eur. J. Oper. Res. **72**(3), 558–573 (1994)
139. Kouvelis, P., Papanicolaou, V.: Expected travel time and optimal boundary formulas for a two-class-based automated storage/retrieval system. Int. J. Prod. Res. **33**(10), 2889–2905 (1995)
140. Kozajda, E.: Logistic indicators for assessing the effectiveness of a storage subsystem in a production company. Scientific Papers of Poznan University of Technology, Organization and Management **46**, 111–126 (2007)
141. Krasucki, J., Rostkowski, A.: The concept of electric-drive power hydraulics systems on the example of a vehicle operating mechanisms MONTRAKS. Mechanical Overview 9/2005, Warsaw (2005)
142. Krawczyk, S., Zajac, P. (eds.): Logistics—Theory and Practice. PUBLISHING DIFIN, Warsaw (2010)
143. Krawczyk, S.: Quantitative Methods in Logistics. Publishing C. H. Beck, Warsaw (2001)
144. Kruszewski, Z., Michalak, G.: Selected Topics in the Theory of Traffic and the Construction of Farm Vehicles. Warsaw University of Technology Press, Warsaw (1989)
145. Kulczyk, J., Winter, J.: Inland Water Transport. Wroclaw University of Technology Press, Wroclaw (2003)

146. Kulturel, S., Ozdemirel, N.E., Sepii, C., Bozkurt, Z.: Experimental investigation of shared storage assignment policies in automated storage/retrieval systems. IIE Trans. **31**(8), 739–749 (1999)
147. Kwasniowski, S., Zajac, P. (eds.): Automatic Identification in Logistics Systems. Wroclaw University of Technology Press, Wroclaw (2004)
148. Kwasniowski, S., Zajac, M., Zajac, P.: Telematic problems of unmanned vehicles positioning at container terminals and warehouses. In: Jerzy Mikulski (ed.) Transport Systems Telematics 10th Conference, TST 2010, vol. 104, pp. 391–399, Katowice-Ustron, Poland. Springer, Berlin, cop. 2011. ISSN 1865-0929, 20–23 October 2010
149. Kwasniowski, S., Zajac, P.: Possibilities of using photovoltaic cells to power lift trucks in a typical warehouse pallet. Mag. Weighing-Dosing-Packag. **1**, 54–59 (2012)
150. Kwasniowski, S., Zajac, M., Zajac, P.: Telematic problems of positioning of unmanned vehicles at container terminals and warehouse. In: Mikulski, J. (ed.) Transport Systems Telematics, pp. 391–399. Springer, Berlin (2010)
151. Lange-Sadzinska, K., Ziemecka, M.: Przewodnik po EDI. Publisher University of Lodz, Lodz (2000)
152. Larson, N., March, H., Kusiak, A.: A heuristic approach to warehouse layout with class-based storage. IIE Trans. **29**, 337–348 (1997)
153. Lee, H.S.: Performance analysis for automated storage and retrieval systems. IIE Trans. **29**, 15–28 (1997)
154. Lee, M.-K., Elsayed, E.A.: Optimization of warehouse storage capacity under a dedicated storage policy. Int. J. Prod. Res. **43**(9), 1785–1805 (2005)
155. Lee, M.-K., Hwang, H.: An approach in the design of a unit-load automated carousel storage system. Eng. Optim. **13**, 197–210 (1988)
156. Lee, Y.H., Tanchoco, J.M.A., Chun, S.J.: Performance estimation models for AS/ RS with annual sized cells. Int. J. Prod. Res. **37**(18), 4197–4216 (1999)
157. Lerher, T., Iztok, P., Sraml, M., Tollazzi, T.: Travel time models for automated warehouses with aisle transferring storage and retrieval machine. Eur. J. Oper. Res. **205**(3), 571–583 (2010)
158. Leszczynski, J.: Modeling of Transport Systems and Processes. Warsaw University of Technology Publishing House, Warsaw (1999)
159. Levy, J.: The optimal size of a storage facility. Nav. Res. Logistics Quart. **21**, 319–326 (1999)
160. Li, J., Sava, A., Xie, X.: Cutting Edge Production Research of the French Community. Int. J. Prod. Res. **47**(2), 403–414 (2009)
161. Li, Y., Wang, Z.: Research on optimization design method of storage unit in distribution center. Proceedings of the 2nd International Conference on Modelling and Simulation, ICMS 2009, vol. 6, pp. 484–489 (2009)
162. Li, Z., Izumi, T., Zhou, H.: Optimal design of lead for minimizing energy dissipated in a mechatronic system with a ball screw-nut. IEEE International Conference on Mechatronics and Automation, ICMA 2009, pp. 1985–1990 (2009)
163. Linn, R.J., Wysk, R.A.: An expert system framework for automated storage and retrieval system control. Comput. Ind. Eng. **18**(1), 37–48 (1999)
164. Liu, S.-N., Ke, Y.-L., Li, J.-X., Lu, Z.: Optimization for automated warehouse based on scheduling policy. Jisuanji Jicheng Zhizao Xitong/Comput. Integr. Manuf. Syst., CIMS **12** (9), 1438–1443 (2006)
165. Lowe, T.J., Francis, R.L., Reinhardt, E.W.: A greedy network flow algorithm for a warehouse leasing problem. IIE Trans. **11**(3), 170–182 (1979)
166. Luxhoj, J.T., Skarpness, B.O.: A manpower planning model for a distribution center: a case study. Mater. Flow **3**, 251–261 (1986)
167. Makris, P.A., Makri, A.P., Provatidis, C.G.: Energy-Saving Methodology for Material Handling Application. Elsevier, Athens (2005)

168. Makris, P.A., Makri, A.P., Proyatidis, C.G.: Energy-saving methodology for material handling applications. Appl. Energy **83**(10), 116–1124 (2006)
169. Malmborg, C.J., AI-Tassan, K.: An integrated performance model for order picking systems with randomized storage. Appl. Math. Model. **24**(2), 95–111 (2000)
170. Malmborg, C.J., AI-Tassan, K.: Analysis of storage assignment policies in less than unit load warehousing systems. Int. J. Prod. Res. **36**, 3459–3475 (1998)
171. Malmborg, C.J.: An integrated storage system evaluation model. Appl. Math. Model. **20**(5), 359–370 (1996)
172. Malmborg, C.J.: Design optimization models for storage and retrieval systems using rail-guided vehicles. Appl. Math. Model. **27**(12), 929–941 (2003)
173. Malmborg, C.J.: Interleaving models for the analysis of twin shuttle automated storage and retrieval systems. Int. J. Prod. Res. **38**(18), 4599–4610 (2000)
174. Malmborg, C.J.: Rule of thumb heuristics for configuring storage racks in automated storage and retrieval systems design. Int. J. Prod. Res. **39**(3), 511–527 (2001)
175. MAN Nutzfahrzeuge, A.G.: Trucknology Generation S-X. MAN Nutzfahrzeuge S.A., Munchen (2009)
176. Marschalek, J., Iowa, M.S.: Tools and techniques for reduced energy consumption with residential energy system case application by State University, p. 215 (2009)
177. Marsh, W.H.: Elements of block storage design. Int. J. Prod. Res. **17**(4), 377–394 (1979)
178. Marsh, W.H.: Storage system optimization. Production and distribution research center. Int. J. Prod. Res. **21**(2), 163–172 (1983)
179. Marshall, W.S., Hamner, P.: Pallets move the world: the case for developing system—based designs for unit loads. Forest Prod. J. **55**(3), 8–16 (2005)
180. Martin, H., Romisch, P., Weidlich, A.: Materialfluss-technik. Viewegs Fachbucher der Technik, Wiesbaden (2004)
181. Mascarenhas, W.F.: Two aspects of the pallet loading problem. Electron. Notes Discrete Math. **19**, 381–387, June 1, 2005
182. Masoud, M.: Storage System Management Using Reinforcement Learning Techniques and Nonlinear Models, p. 154. Ph.D., University of Waterloo, Canada (2009)
183. McAree, P.W.: Models for the Design and Analysis of a Large Package Sort Facility. Ph.D., University of Maryland College Park (2001)
184. McGinnis, L.F.: Best of breed warehouse performance assessment. Annual Conference Council on Logistics Management, Chicago (2003)
185. Merohatgi, A.: www.ece.gatech.edu/research/UCEP/
186. Meller, R.D., Gau, K.Y.: Performance analysis of split-case sorting systems. Manuf. and Serv. Oper. Manage. **4**(4), 258–274 (2002)
187. Meller, R.D., Mungwattana, A.: Multi-shuttle automated storage/retrieval systems. IIE Trans. **29**(10), 925–938 (1997)
188. Meller, R.D., Klote, J.F.: A throughput model for carousel/VLM. IIE Trans. **36**(8), 725–741 (2004)
189. Michel, J.C.F., Millner, H., Yossiek, M.: Positioning a novel wireless forklift positioning system for indoor and outdoor. Navig. Commun. **2008**, 219–227, Jan 1, 2008
190. Min, H.: The applications of warehouse management systems: an exploratory. Int. J. Logistics Res. Appl. Lead. J. Supply Chain Manage. **9**(2), 111–126, Jan 1, 1469
191. Minav, T.A., Laurila, L.I.E., Immonen, P.A., Haapala, M.E., Pyrhonen, J.J.: Electric energy recovery system efficiency in a hydraulic forklift. EUROCON2009, EUROCON '09, Issue 10, pp. 758–765, Jan 1, 1109
192. Moder, J.J., Thornton, H.M.: Quantitative analysis of the factors affecting floor space utilization of palletized storage. J. Ind. Eng. **16**(1), 818–832 (1995)
193. Moleeratanond, W., Kramer, A., Ashby, B.H., Bailey, W.A., Bennett, A.H.: Effect of temperature fluctuations on energy consumption and quality changes of palletized foods in frozen storage. ASHRAE Trans. **85**(2), 56–65 (1999)

194. Morabito, R., Morales, S.R., Widmer, J.A.: Loading optimization of palletized products on trucks. Transp. Res. Part E, Logistics Transp. Rev. **36**(4), 285–296 (2000)
195. Mrowczynska, B.: An application of evolutionary and immune algorithms for the optimization of packing a diversified set of packets on a pallet. Maintenance Probl. **4**, 137–145 (2008)
196. Muller, D.J.: AS/RS and warehouse modeling. Winter Simulation Conference Proceedings, pp. 802–814 (1998)
197. New concepts in storage logistics—paper industry opts for automation (New concepts in storage logistics—paper industry opts for automation) Anon Source: Internationale Papierwirtschaft IPW, **8**, 21–22 (2000)
198. Nowakowski, T.: Reliability Logistics Systems. Wroclaw University of Technology Press, Wroclaw (2011)
199. Nowicka-Skowron, M.: Efficiency of Logistics Systems. Polish Economic Publishing House, Warsaw (2000)
200. Pan, C.-H., Wang, C.-H.: A framework for the dual command cycle travel time model in automated warehousing systems. Int. J. Prod. Res. **34**(8), 2099–2117 (1996)
201. Pandit, R., Palekar, U.: Response time considerations for optimal warehouse layout design. J. Eng. Ind, pp. 322–328 (1993)
202. Park, B.C., Foley, R.D., White, J.A., Frazelle, E.H.: Dual command travel times and mini load system throughput with turnover-based storage. IIE Trans. **35**, 343–355 (2003)
203. Park, B.C., Frazelle, E.H., White, J.A.: Buffer sizing models for end-of-aisle order picking systems. IIE Trans. **31**, 31–38 (1993)
204. Park, B.C., Park, J.Y., Foley, R.D.: Carousel system performance. J. Appl. Probab. **40**, 602–612 (2003)
205. Park, Y.H., Webster, D.B.: Modelling of three-dimensional warehouse systems. Int. J. Prod. Res. **27**(6), 985–1003 (1989)
206. Park, Y.H., Webster, D.B.: Modelling of three-dimensional warehouse systems. Int. J. Prod. Res. **27**(6), 985–1003 (1999)
207. Paschalidis, I.C., Li, K., Estanjini, R.M., Lin, Y., Guok, D.: Intelligent forklift dispatching in warehouses using a sensor network. Control Autom. **09**, 112–114, Jan 1, 2009
208. Pedersen, H., Hansen, M., Andersen, T., Lindholdt, P.: An optimization approach applied to design the hydraulic power supply for a forklift truck. Am. Soc. Mech. Eng. The Fluid Power and Systems Technology Division (Publication) FPST, vol. 11, pp. 189–196 (2004)
209. Perlmann, A.M., Bailey, M.: Warehouse logistic systems—a CAP model. Eng. Costs Prod. Econ. **13**(3), 229–237 (1988)
210. Perlmann, A.M., Bailey, M.: Warehouse logistics systems—A CAD model. Eng. Costs Prod. Econ. **13**, 229–237 (1989)
211. Petersen, C.G.: An evaluation of order picking policies for main order companies. Prod. Oper. Manage. **9**(4), 319–335 (2000)
212. Phillips, E., Quarterman, L.: Warehouse and layout planning. Department of the Navy, Naval Supply Systems Command NAVSUP Publication 529 (1985)
213. Piatkiewicz, A., Sobolski, R.: Cranes. Publishing WNT, Warsaw (1969)
214. Pliskin, J.S., Dori, D.: Ranking alternative warehouse area assignments: a multiattribute approach. IIE Trans. **14**(1), 19–26 (1982)
215. Pluta, Z.: Solar Energy Installations. Warsaw University of Technology Publishing House, Warsaw (2008)
216. Poess, M., Othayoth, N.R.: A power consumption analysis of decision support systems. WOSP/SIPEW'10—Proceedings of the 1st Joint WOSP/SIPEW International Conference on Performance Engineering, pp. 147–152, 2010, Modern Materials Handling, vol. 61(13), pp. 49–52, December 2006
217. Potrc, I., Lerher, T., Kramberger, J., Sraml, M.: Simulation model of multi-shuttle automated storage and retrieval systems. J. Mater. Process. Technol. **157**, 236–244 (2004)
218. Collective work: Mechanics Explained. Publishing WNT, Warsaw (1994)

219. Randhawa, S.U., McDowell, E.D., Wang, W.-T.: Evaluation of scheduling rules for single- and dual-dock automated storage/retrieval system. Comput. Ind. Eng. **20**(4), 401–410 (1991)
220. Randhawa, S.U., Shroff, R.: Simulation-based design evaluation of unit load automated storage/retrieval systems. Comput. Ind. Eng. **28** (1), 71–79 (1995)
221. Rao, A.K., Rao, M.R.: Solution procedures for sizing of warehouses. Eur. J. Oper. Res. **108**, 16–25 (1998)
222. Rabkowski, J.: Energy storage interfaces with Z-source based inverters. Arch. Electr. Eng. **58** (3–4), 143–156 (2009)
223. Recknagel, H., Ginsberg, O., Gehrensberg, K., Sprenger, E., Honmann, W.: Taschenbuch fur Heinzung und Klimatechnik. Ernst-Rudolf Schramek Uniwersytet Dortmund, Munchen (2008)
224. Rizzi, A., Zamboni, R.: Efficiency improvement in manual warehouses through ERP systems implementation and redesign of the logistics processes. Logistics Inf. Manage. **12**(5), Jan 1, 1999
225. Roberts, S.D., Reed, R.: Optimal warehouse bay configurations. AIIE Trans. **4**(3), 178–185 (1972)
226. Roli, Y., Rosenblatt, M.J., Kadosh, D.: Determining the size of a warehouse container. Int. J. Prod. Res. **27**(10), 1693–1704 (1998)
227. Roli, Y., Rosenblatt, M.J.: Random versus grouped storage policies and their effect on warehouse capacity. Mater. Flow **1**, 199–205 (1993)
228. Roodbergen, K.J., Vis, I.F.A.: A model for warehouse layout. IIE Trans. **38**(10), 799–811 (2006)
229. Rosen, M.A.: Exergy in industry: accepted or not? Exergy Int. J. **1**(2), 67 (2001)
230. Rosenblatt, M.J., Eynan, A.: Deriying the optimal boundaries for class-based automatic storage/retrieyal systems. Manage. Sci. **35**(12), 1519–1524 (1989)
231. Rosenblatt, M.J., Roli, Y., Zyser, V.: A combined optimization and simulation approach for designing automated storage/retrieyal systems. IIE Trans. **25**(1), 40–50 (1993)
232. Rosenblatt, M.J., Roli, Y.: Warehouse design with storage policy considerations. Int. J. Prod. Res. **22**(5), 809–821 (1993)
233. Rosenblatt, M.J., Roli, Y.: Warehouse capacity in a stochastic environment. Int. J. Prod. Res. **26**(12), 1847–1851 (1998)
234. Ross, H.J. (ed.).: Aufbau einer Logistik-Ausbildung in Polen. TEMPUS JEP-03238 Band I, Universitat Stuttgart, Stuttgart (1992–1995)
235. Ross, H.J. (ed.).: Aufbau einer Logistik-Ausbildung in Polen. TEMPUS JEP-03238 Band II, Universitat Stuttgart, Stuttgart (1992–1995)
236. Ross, H.J. (ed.).: Aufbau einer Logistik-Ausbildung in Polen. TEMPUS JEP-03238 Band III, Universitat Stuttgart, Stuttgart (1992–1995)
237. Ross, A., Droge, C.: An integrated benchmarking approach to distribution center performance using DEA modeling. J. Oper. Manage. **20**, 19–32 (2002)
238. Rowenhorst, B., Reuter, B., Stockrahm, V., van Houtum, G.J., Mantel, R.J., Zijm, W.H.M.: Warehouse design and control: framework and literature review. Eur. J. Oper. Res. **122**, 515–533 (2000)
239. Russ, A.: Robot warehouse. Rob. Age **6**(4), 27–28 (1984)
240. Russell, M.L., Meller, R.D.: Cost and throughput modeling of manual and automated order fulfillment systems. IIE Trans. **35**, 589–603 (2003)
241. Russell, T.D., Malstrom, E.M.J., Meeks, H.D.: Two-dimensional palletizing procedure for warehouse loading operations. IIE Trans. (Inst. Ind. Eng.) **20**(4), 418–425 (1988)
242. Rutkowski, K. (ed.): Logistics Distribution. PUBLISHING DIFIN, Warsaw (2000)
243. Saga Andrew, P., Rouse, W.B.: Handbook of Systems Engineering and Management, 2nd edn. Wiley, Hoboken (2009)
244. Sakamoto, A., Tsukiyama, S., Shimizu, K.: A consideration on a studying palette problem in a two-dimensional automatic warehouse. Proceedings—IEEE International Symposium on Circuits and Systems, vol. 4, pp. 2849–2852 (1990)

245. Salvendy, G.: Handbook of Industrial Engineering—Technology and Operations Management, 3rd edn. Wiley, Hoboken (2001)
246. Schefczyk, M.: Industrial benchmarking: a case study of performance analysis techniques. Int. J. Prod. Econ. **32**, 1–11 (1993)
247. Schwarz, L.B., Graves, S.C., Hausman, W.H.: Scheduling policies for automatic warehousing systems: simulation results. AIIE Trans. **10**(3), 260–270 (1998)
248. Seelinger, M., Yoder, J.D.: Automatic pallet engagement by a vision guided forklift. Robot. Autom. 4068–4073, Jan 1, 2005
249. Sen, Z.: Solar energy in progress and future research trends. Prog. Energy Combust. Sci. **30**, 367–416. www.sciencedirect.com (2004)
250. Lu, S., Wu, Y., Fu, Y.: Research and design on pallet-throughout system based on RFID. IEEE International Conference on Automation and Logistics, **1109**(10), 2592–2595, Jan 1, 2007
251. Sharp, G.P., Vlasta, D.A., Houmas, C.G.: Economics of storage/retrieyal systems for item picking. Material Handling Research Center, Georgia Institute of Technology, Atlanta (1994)
252. Shugin, W., Lindu, Z.: Optimization of goods location numbering and storage and retrieval sequence in automated warehouse. Proceedings of the 2009 International Joint Conference on Computational Sciences and Optimization, CSO 2009, vol. 2, pp. 883–886 (2009)
253. Silva, C.A., Sousa, J.M.C., Runkler, T.A.: Optimization of logistic system using fuzzy weighed aggregation. Fuzzy Sets Syst. **158**(77), 1947–1960 (2007)
254. Silka, W.: Energy Intensity of the Car Traffic. WNT, Warsaw (1997)
255. Soe, S.Y., Hong, H.K., Hye, J.J.: Development of an Air Force warehouse logistics. Eur. J. Oper. Res. **183**(1), 148–161 (2007)
256. Special issue on emergency logistics management transportation research, Part E: Logistics and Transportation Review Transportation, Research Part E **41**(5), 459–460 (September, 2005)
257. Speranza, M.G., Ukovich, W.: Analysis and integration of optimization models for logistic systems. Int. J. Prod. Econ. **35**(1–3), 183–190 (1994)
258. Spieker, S., Rohrig, C.: Localization of pallets in warehouses using wireless sensor networks. 16th Mediterranean Conference on Control and Automation, **1109**(10), 1833–1838, Jan 1, 2008
259. Stachura, A.: The laboratory model of hybrid propulsion system of heavy duty forklift. Modelling and Optimization Physical System, Silesian University of Technology, Gliwice (2009)
260. Stolze, C.: Pallet handling with robotrailers in a large automated distribution center. Conference: Proceedings of the Ist International Conference on Automated Guided Vehicle Systems, IFS (Publ) Ltd, pp. 27–34 (1981)
261. Su, C.-T.: Performance evaluation of carousel operation. Prod. Planning Control **9**(5), 477–488 (1998)
262. Sulima, K.: Analysis of technical standards and regulations in relation to the lift pallet trucks. Promoter Kosiara A., Paper is not published, Wroclaw University of Technology, Wroclaw (2008)
263. Sybil, S.: Integrated warehouse system. Conference: SME, April 7, 1995–April, No 10, 1995
264. Szargut, J., Petela, R.: Egzergia. Publishing WNT, Warsaw (1965)
265. Szumanowski, A.: The theory of the car. The accumulation of energy in vehicle. Publishing and Communications and Communications, Warsaw (1984)
266. Thonemann, U.W., Brandeau, M.L.: Optimal storage assignment policies for automated storage and retrieval systems with stochastic demands. Manage. Sci. **44**(1), 142–148 (1998)
267. Tsai, R., Malstrom, E., Meeks, H.: Robotic utilization of warehouse pallet loads. Proceedings —Fali Industrial Engineering Conference (Institute of Industrial Engineers), pp. 227–238 (1985)

268. Tunel, I., Craig, P., Levine, M., McMahon, J., McCollister, G., Hesterberg, B., Robinson, M.: Estimation of energy intensity by end-use for commercial buildings. Energy **12**(6), 435–446 (1997)
269. Twarog, J.: Counters and Indicators Logistics. Publishing ILiM, Poznan (2003)
270. Ullrich, H.-J.: Cooling Technique, vol. 1. Publishing "IPPU Masta", Gdansk (1998)
271. Ullrich, H.-J.: Cooling Technique, vol. 2. Publishing "IPPU Masta", Gdansk (1999)
272. van Oudheusden, D.L., Tzen, Y.-J., Ko, H.-T.: Improving storage and order picking in a person-on-board AS/R system. Eng. Costs Prod. Econ. **13**, 273–283 (1998)
273. Wall, G., Gong, M.: On exergy and sustainable development-part 2: indicators and methods. Exergy Int. J. **1**, 217–233 (2001)
274. Wall, G., Gong, M.: On exergy and sustainable development-part 1: conditions and concepts. Exergy Int. J. **1**, 128–145 (2001)
275. Wang, J.-Y., Yih, Y.: Using neural networks to select a control strategy for automated storage and retrieval systems (AS/RS). Int. J. Comput. Integr. Manuf. **10**(6), 487–495 (1997)
276. Wasiak, M.: A guiding theory approach to logistics systems modelling. Arch. Transp. **19**(3), 103–120 (2007)
277. Weber, R.C., Gies, P., Seifert, J.: Energy conservation on forklift truck operations. Gateway Energy Conference, pp. 554–558 (1998)
278. White, J.A., Francis, R.L.: Normative models for some warehouse sizing problems. AIIE Trans. **9**(3), 185–190 (1971)
279. White, J.A., DeMars, N.A., Matson, J.O.: Optimizing storage system selection. In: Proceedings of the 4th International Conference on Automation in Warehousing, Tokyo, Japan (1991)
280. Wnuk, R., Berent-Kowalska, G., Peryt, Sz.: Energy efficiency in the years 1995–2005. Central Statistical Office, Department of Economic Statistics, National Agency for Energy Conservation (1995–2005)
281. Wood, D.R., Barone, A.R., Murphy, P.R., Wardlow, D.L.: Book Division of American Management Association, AMACOM (2002)
282. Wu, F., Lu, M.-C.: Modeling of warehouse and its analyses. Proceedings of the International Conference on Modeling, Simulation and Visualization Methods, MSV04, pp. 239–245 (2004)
283. Wurll, Ch.: Pick to pallet, automated order picking with industrial robots. VDI Berichte **1679**, 167–172 (2002)
284. www.binnenvaart.be
285. www.ecotransit.org
286. www.elogistics101.com
287. www.fotowoltaika.net
288. www.jungheinrich.com
289. www.markt-intern.com
290. www.muratorplus.pl
291. www.nau-gmbh.de
292. www.ptpiree.pl
293. Yahia, E.M.: Modified and Controlled Atmospheres for the Storage of Horticultural Commodities. Taylor & Francis Group Ltd, USA (2009)
294. Yan den Berg, J.P.: Survey on planning and control of warehousing systems. IIE Trans. **31**, 751–762 (1999)
295. Yanek, F.M., Albright, L.D.: Energy systems engineering—evaluation and Implementation. www.springer.com (2003)
296. Yoon, C.S., Sharp, G.P.: A structured procedure for analysis and design of order pick systems. IIE Trans. **28**, 379–389 (1996)
297. Yoon, C.S., Sharp, G.P.: Case application of the cognitive design procedure for an order pick system. Case study. Eur. J. Oper. Res. **87**, 223–246 (1998)

298. Zajac, P. (ed.): Logistics Fleet Management Road. Wroclaw University of Technology Press, Wroclaw (2003)
299. Zajac, P.: A practical method of determining the energy intensity in the service of a loading unit in a warehouse. In: Grzybowska, K. (ed.) Logistics: Selected Concepts and Best Practices: Monograph, pp. 235–262. Publishing House of Poznan University of Technology, Poznan (2012). (Logistics & Production)
300. Zajac, P.: Can the raising of energy consumption of information interchange be a factor that reduces the total energy consumption of logistic warehouse system? In: Chlebus, E. (ed.) Production Engineering: Innovations and Technologies of the Future. Institute of Production Engineering and Automation, Wroclaw University of Technology, Wroclaw, pp. 79–89. International Conference Production Engineering 2011, Wroclaw, 30 June–1 July 2011
301. Zajac, P.: CRM—Customer Relationship Management in Distribution Logistics. Wroclaw University of Technology Press, Wroclaw (2007)
302. Zajac, P.: Electronic Exchange of Data in Logistic Systems. Wroclaw University of Technology Press, Wroclaw (2010)
303. Zajac, P.: Evaluation Study of Enterprise Delivery Chain Logistics Support Systems Functional Features Supplement. Polish Systems Society, Wroclaw University of Technology Press, Wroclaw (2003)
304. Zajac, P.: The concept of system evaluation model transportation and storage. Total Logistic Management, TLM 2009 [e-dokument]: XIII Conference on Applied Logistics, Zakopane. Felix, G. (ed.) Proceedings of Transport Committee of the Polish Academy of Sciences, pp. 25–28 November 2009
305. Zajac, P.: The method of reducing energy consumption forklift working. Logistics Mag. **6**, 4057–4065 (2011)
306. Zajac, P.: Wheelchair sampling methodology DIS-2 with regard to energy intensity. Logistics Mag. **3**, 3065–3074 (2011)
307. Zajac, P.: Evaluation of computer systems supporting logistics storage systems. Syst.-J. Transdisciplinary Syst. Sci. **16**(2), 423–438 (2012)
308. Zajac, P.: The issue of determining cycles for mobile storage devices on the example of forklift. Logistics Mag. **6**, 4047–4055 (2011)
309. Zajac, P.: Information systems and telematics in logistics. In: Krawczyk, S. (ed.) Logistics: Theory and Practice, vol. 2, pp. 115–144. Publishing DIFIN, Warsaw (2011)
310. Zajac, P.: Storage Systems, a Series of Edulog. Publishing WDiO, Wroclaw (2010)
311. Zajac, P.: The choice of parameters of logistic warehouse system, with taking the energy into consideration. In: Grzybowska, K., Golinska, P. (eds.) Selected Logistics Problems and Solutions: Monograph, pp. 107–120. Publishing House of Poznan University of Technology, Poznan (2011). (Logistics & Production)
312. Zajac, P.: The idea of the model of evaluation of logistics warehouse systems with taking their energy consumption under consideration. Arch. Civil Mech. Eng. **11**(2), 479–492 (2011)
313. Zajac, P.: Transport-storage system optimization in terms of exergy. In: Fertsch, M., Grzybowska, K., Stachowiak, A. (eds.) Modelling of Modern Logistics Enterprises: Monograph, pp. 71–83. Publishing House of Poznan University of Technology, Poznan (2009)
314. Zajac, P.: Transport-storage system optimization in terms of exergy. In: Agostino, Bruzzone (ed.) The 13th International Conference on Harbor Maritime Multimodal Logistics Modeling and Simulation, HMS 2010. Fes, Morocco, Laboratories des Sciences de l'Information et des Systemes, Marseille, cop. 2010, pp. 143–148, October 13–15 2010
315. Zekai, S.: Solar energy in progress and future research trends. Prog. Energy Combust. Sci. **30**, 367–416 (2004)
316. Zembrzuski, K.: The Theory of Propulsion and Braking of the Train. Publishing PWN, Warsaw (1978)

317. Zeng, A.Z., Mahan, M., Fluet, N.: Designing an efficient warehouse layout to facilitate the order-filling process: an industrial distributor's experience. Prod. Inventory Manage. J. **43**(3–4), 83–88 (2002)
318. Qinghua, Z., Jun, W., Guoan, C., Zhuan, W., Dawei, Y., Shanshan, Z.: Pallet rental information system based on RFID. Ind. Electron. Appl., pp. 886–891, Jan 1, 2009
319. Zhiyi, L.: Grey correlation analysis on logistics energy consumption. Proceedings of the 9th International Conference of Chinese Transportation Professionals, ICCTP 2009: Critical Issues in Transportation System Planning, Development, and Management, vol. 358, pp. 3216–3221 (2009)
320. Zollinger, H.A.: Expanded methodology to concept horizontal transportation problem solutions. In: Graves, R.J., McGinnis, L.F, Medeiros, D.J., Ward, R.E., Wilhelm, M.R. (eds.) Progress in Material Handling Research, pp. 651–663 (2002)
321. Zoran, R., Yladeta, C., Zlatko, H.: Some aspects of storage and bulk queueing systems in transport operations. Transp. Plann. Technol. **20**(1), 67–81, Jan 1, 1996

MIX
Papier aus verantwortungsvollen Quellen
Paper from responsible sources
FSC® C105338

If you have any concerns about our products,
you can contact us on
ProductSafety@springernature.com

In case Publisher is established outside the EU,
the EU authorized representative is:
Springer Nature Customer Service Center GmbH
Europaplatz 3, 69115 Heidelberg, Germany

Printed by Libri Plureos GmbH
in Hamburg, Germany